AIR AMERICA

CIA SUPER PILOT SPILLS THE BEANS!

FLYING HELICOPTERS IN LAOS FOR AIR AMERICA

BY CAPTAIN BILL COLLIER
AIR AMERICA H-34 AND BELL HUEY PILOT

Praises for this book

FIRST PLACE WINNER!

This book won First Place
in the 2016
Idaho Writers League
annual state-wide writing contest,
"book, non-fiction" category.

I was [radio] witness to Captain Bill's death-defying effort to rescue an Air Force Raven shot down on the Plain of Jars. He displayed Medal-of-Honor guts. Read this book and you'll know much more about the day-to-day combat and courage required of those special few who fought the communists in Laos.

Jim "Mule" Parker, CIA Officer

This book is an intriguing look into the daily lives of the pilots who flew for the CIA in Southeast Asia during the Vietnam War. Collier pulls no punches. He describes it so well that you can smell burning fuel, hear the roar of the chopper's engines, feel the bullets passing by, and taste the food and cans of French Beaujolais. Read it! You will not be disappointed!

Mark V. Garrison, U.S. Army gunship helicopter pilot in Vietnam.
author of the number one bestseller, *Guts 'N Gunships.*

Air America, the CIA's proprietary air arm, provided covert logistic and combat support to our indigenous allies in Asia. Captain Bill Collier tells a fascinating story as he takes the reader on a day-by-day tour through the Neverland of clandestine warfare in Southeast Asia. His book offers rare insight into an intriguing chapter in American History.

Marion Sturkey, author of *Bonnie-Sue, No Greater Love, Warrior Culture of the U.S. Marines*, and others.

In this exciting follow-up to his first memoir (*The Adventures of a Helicopter Pilot)*, Captain Collier steps into the secret business of the CIA in Laos. He tells of countless near misses and daring flights, while also giving a peek at the raucous life of the pilots on the ground. More than a flying memoir, this book follows a group of close-knit, daring, young men as they carouse their way around Southeast Asia. Both a sweeping view of CIA activities in the early seventies and an intimate portrait of one man's life, "CIA Super Pilot" is a must-have for anyone interested in flying, history, or the human spirit.

April Davila, MS in Creative Writing, University of Southern California.
Soon-to-be world-famous novelist and mother of two
of Captain Collier's most wonderful grandchildren.

This is a five-star account of the real story of Air America and the men who lived it. Factual and much more exciting than the mortally flawed Hollywood version, it provides a candid, in-depth look at clandestine CIA air operations in Southeast Asia during the Vietnam conflict. From the 'Secret CIA Base at Long Tieng' to his base at Udorn, Thailand and on down to the fancy hotels of Bangkok and Hong Kong, Bill takes the reader on most exciting adventures in H-34 Helicopters and Bell Hueys, too.

Rich Faletto, Major USAF, Ret.
Former "Owner" of CIA Lima Sites
Udorn Royal Thai Air Force Base, 1973

Collier's stories exemplify the adrenaline rush inherent in flying helicopters. He recounts plenty of excitement from everyday life, too. His many adventures—both in and out of air machines—are wonderfully readable.

Lt. Col. Henry Zeybel, USAF (Retired)

This book follows closely on the heels of his earlier *Adventures of a Helicopter Pilot* and is every bit as readable. The chapters alternate between stories of his hair-raising exploits while flying for Air America and tales of pilots' adventures when not in the sky. There are tears and laughter for every reader. Captain Collier has a superior vocabulary, which emphasizes the emotional intensity of his experiences. While the title is certainly apt, I found myself wanting to rename it something like *Guys and Dolls* and *Oh, shit, we're gonna die!* However, such a title would not do justice to the death-defying risks taken by these pilots, and often for purposes they themselves did not understand. As the wife of a former Air America captain, I found the book well structured, carefully thought out, and fully in keeping with stories of Air America that I have heard from my spouse.

Sylvia C. Davis, PhD, professor emerita of German and Humanities,
Eastern Kentucky University

I found the book engrossing from cover to cover. While I really don't question any of the flying episodes, some of the sexual exploits make me wonder if I spent too much time flying and not enough around the officers' clubs and bars in certain upscale (and some not-so-upscale) hotels. Captain Collier's text brings back many memories of adventures I had and people I knew. The stories will appeal to a wide audience, from those who lived some of these same adventures to neophytes in the world of aviation.

O.L. ("Lee") Howell, Captain, Air America

From Viet Nam war hero to CIA pilot, you'll love Bill's wild-ride adventures as he opens his kimono and tells all!

Suzen Fiskin, Inspirational speaker
Author of Playboy Mansion Memoirs.

Vietnam War correspondent, Anne Darling, visited the Air America helicopter base in Udorn, Thailand. She hung out with the fellows and pumped them for information. She then wrote an article for the premiere issue of **oui,** an offshoot of PLAYBOY Magazine. The article was entitled:

"CIA Super Pilots Spill the Beans!"

I was one of those Super Pilots. This is my story.

(The entire article is recreated in its entirety, with no word changes, in Appendix A, with express permission of the owner of **oui** magazine.)

Disclaimers

I reveal NO national secrets in this book.

I expose NO CIA secret methods or techniques.

AIR AMERICA
A CIA SUPER PILOT SPILLS THE BEANS!

FLYING HELICOPTERS IN LAOS FOR AIR AMERICA

Photo by Gary Connolly

BY CAPTAIN BILL COLLIER

AIR AMERICA H-34 AND BELL HUEY PILOT

Second Edition

Cover photo taken early 1994 by self-pro-claimed, "World's Greatest Helicopter Pilot" Andy Campbell. Helicopter in photo actually was Evergreen Helicopter Company's Bell 205A-1, N4750R. Photo highly modified to look like Air America Bell 205 XW-PFH. Cargo netting taken from Air America Newsletter "Vol. IV, No. 1, OKINAWA, 1972." Water buffalo head found on Wikipedia.

Title page photo taken by Air America Captain Gary Connolly on the rooftop of the Chai Porn Hotel, Udorn, Thailand.

Back Cover: Cover of oui magazine used with permission of **oui** magazine owner. Picture of author on the back page is from author's 1972 Passport. Photo of Long Tieng courtesy of Peter and Baythong Whittlesey, authors of *Sinxay: Renaissance of a Lao-an Epic Hero.*

Most photographs and illustrations in this publication were taken from the author's personal archives. Those that were not are appropriately credited to their source.

For a signed copy send $20.00 plus $5.00 for shipping and handling to:
WANDERING STAR PRESS
P.O. Box 651
Sandpoint, Idaho 83864

Collier, Bill

AIR AMERICA, A CIA Super Pilot Spills the Beans
Flying Helicopters in Laos for AIR AMERICA

Second Edition
ISBN 978-1688081871
L.O.C. No.: 2017912204

Contents

Praises for this Book iii
Disclaimers vii
Title Page .....
Copyright page x
Indexed Map of Laos xvi
DEDICATION to Gary Connolly xviii
PREFACE – Recalled! xxiii
1. Worry One – A Near Fiery Crash 1
2. Worry Two – Seriously Breaking the 12-hour Rule 7
3. Routine Work 11
4. Black Cloud at Udorn – A "Special" Too Quiet! 14
5. First STO – Stalking Susan 21
6. Worry Three – I Insult the Base Vice President 25
7. Background 27
8. Off On a Grand Adventure. 33
9. Settling into Udorn, Thailand 38
10. Training 40
11. I Check Out as Captain. 42
12. Second STO – Chang Mai, Thailand 46
13. Dangerous Uniforms. 48
14. Savannakhet (SKT) 49
Gecko Racing 49
Friendly Fire 50
15. Strawberry Ice Cream 52
16. Bangkok 59
17. Cracked Impeller Housing 64
18. We Move Out of Chai Porn Hotel to Better Quarters 71
Air America Pilot Beats Up Thief 73
FuckYou Lizards 74
19. STO, Beirut 75

20. Multiple malfunctions 79
21. STO June 1971 86
22. Collective Jam-up 89
23. My Longest Day – Almost stranded overnight 91
24. Bizarre Cargoes 95
25. Our In-House Intelligence 98
26. About Drug Running 100
Three Drug Runners 102
27. The Movie AIR AMERICA 103
28. Tony Poe, LS-118A 106
29. We Move Again 108
30. First Annual Leave – A Trip Around the World 111
Tel Aviv, Israel 111
Istanbul, Turkey 112
Greece 112
England 112
31. Engine Cough with Bob Caron before a Dangerous Special 115
Prostitutes for the Local Troops 116
32. A Heavy Lift Job 117
33. BIM Blow Out 119
34. Staying Overnight at Long Tieng 122
35. The Battle for Skyline Ridge 125
36. Another Heavy Lift Job 131
Ms. Ott's Engagement Ring 132
37. Our Base at Ban Houie Xai 135
38. Life in Thailand 138
39. Christmas Fireworks 140
Two Days of Crane Chase. 141
SAR for Royal Lao Army H-34 near Pakxe 143
Twelve Specials Near Louang Prabang 144
40. Worry Four 147
The Rescue Of Raven 1-1 147
We Rescue a Lao pilot. 159

41. Worry Five – Drinking Wine on Duty 160
42. Dual Porter Crash at LS-69A 164
43. Worry Six 166
44. Another STO – The Taj Mahal and Nepal with Sweet Syn..... 169
45. Swamped! 172
46. Henry's Roadside Bar and Grill, Pakxe, Laos. 174
A Round of BJs. 175
Another BJ 176
47. A Disastrous Special Mission – A Crewman Dies 177
48. More Cheap Thrills – I Lift a Heavy TACAN Box at LS-272 ... 181
49. The Press 183
50. Worry Seven – I Insult Two Very Senior Pilots 185
Worry Eight All the above? 186
51. Sleeping in the Cockpit While Flying 187
52. Special Flight into Cambodia 189
Round Engines 191
53. A Newer, Much Bigger Worry 192
54. Close Calls 196
55. Bell Hueys 199
56. July 1972 STO – Hong Kong 207
57. Black Ops – Silent Helicopters 210
58. About Parachutes 214
59. STO to Australia 217
60. I Carry a lot of Baggage 220
61. Second Annual Leave 222
Trattorias 224
62. Working Bells at Long Tieng 228
63. An STO to Hong Kong 232
Submariners at the Hilton Bar 232
64. My Scariest Mission Ever! 236
65. Laced at Long Tieng – Customer Wounded 242

66. The Beginning of the End....246
Last STO....246

67. Lightning Love 248

68. *Mak Mak* DANGER! – I Resign....251

69. Departing Thailand 258

70. More Follow-up Stories....261
One - My First Air America Reunion 261
Two - A Never Before-met Friend from the Far Past....263
Three - 1994 Another Midnight Rambler....264
Four - A Funeral....265
Five - Madras, Oregon....267
Six - A Final Salute to Old Friends. 269
Seven - The Book that Will Never be Written....271
Eight -About Rescues 271
Nine – Awards 272

APPENDIX A **oui** Magazine Article in Full 274
APPENDIX B LOOKEAST Magazine Article about Hill Tribes.. 291
Glossary 297

BIBLIOGRAPHY 300
Videos 301
Web Sites 302
Acknowledgements. 303
About the Author 306

A B C D E F G H I J K L M N O P Q

1 2 3 4 5 6 7 8 9 10 11 12 13

Mengzi
Xin'ansuo
Tongguan
Daxingzhai
Honghe
Pu'er
Lancang
Simao
Dayakou
Sichuanzhai
Jiangxizhai
Pinghe
Jiangcheng
Jinping
Hekou
Lao Cai
Sa Pa
Mengwang
Mandun
Dadugang
Jinghong
Yiwu
Menghai
Menghun
Mengzhe
Muang Ou Nua
Sop Pong
Muang Ou Tai
Muang Va
Ngay Nua
Phôngsali
Boun Nua
Muang Yo
Mengla
CHINA
LAOS
Muang Sing
Louang Namtha
Muang Long
Muang Hay
Muang La
Muang Xay
Viangphoukha
Muang Meung
Muang Bèng
Ban Namnga
Muang Sung
Ban Kheun
Muang Houn
Muang Paktha
Muang Pakbèng
Mekong
Khong
Chiang Rai
Ban Thanoun
Louangphrabang
Ban Samang
Ban Xenkhalok
PLATEAU DE
XIANGKHOANG
Ban San X
Muang Hongsa
Muang Thadua
Muang Xaignabouri
Ban Pakneun
Muang Phoun
Ban Thieng
Muang Phiang
Muang Vangviang
Ban Nale
Ban Hin Heup
Muang Liap
Muang Saiapoun
Muang Khi
Tourakom
Ban Donhiang
Ban Hetkiang
Muang Pak-Lay
Ban Nam Tan
Viangchan
Ban Naxon
Ban Thabok
Nam Ngum Res.
Muang Soum
Ban Tian Sa
Ban Sam Pong
Muang Pakxan
Ban Namcha
Phou Bia 2819
Xiangkhoang
Thun Chang
Nan
Sa
Ban Salik
Na Noi
Phrae
Sirikit Res.
Tha Pla
THAILAND
LUANG PRABANG RANGE
Muang Ngoy
LAO
Muang Khoua
Pak Ban
Dien Bien Phu
Son La
Tuan Giao
Tua Chau
Than Uyen
Lai Chau
Muong Te
Sinh Ho
Fan Si Pan 3143
Phong Tho
Bao Ha
Bac Quang
Ha Giang
Maguan
Matizhen
VIETNAM
Dong Van
Bao Lac
Soc Giang
Cao Bang
Dong Khe
That Khe
Bac Can
Binh Gia
Dinh Ca
Lang Son
Pingxiang
Ningming
Mubian
Dizhou
Jingxi
Tuyen Quang
Cho Chu
Binh Ca
Yen Bai
Nghia Lo
Phu Tho
Viet Tri
Phu Yen
Na San
Ta Khoa
Yen Chau
Moc Chau
Son Tay
Ha Dong
Hoa Binh
Thai Nguyen
Nha Nam
Bac Giang
Bac Ninh
Ha Noi
Hai Duong
Hai Phong
Dong Trieu
Hung Yen
Thai Binh
Do Son
DAO CAT BA
Hong Gai
Cam Pha
Dinh Lap
Tien Yen
Phu Ly
Nam Dinh
Ninh Binh
Phat Diem
Hoi Xuan
Muong Het
Sop Hao
Xam Nua
Ban Ngam
Bai Thuong
Thanh Hoa
Sam Son
Muong Hinh
Qui Chau
Nghia Dan
Tuong Duong
Khe Bo
Con Cuong
Phu Huu
Cua Lo
Vinh
Ben Thuy
Long Moc
Linh Cam
Ha Tinh
Ha Tan
Cam Xuyen
Huong Khe
Ky Anh
Tuyen Hoa
Ron
Ban Soppheung
Ban Naxouang
Ban Songkhon
Ban Nam Chan
Phon Phisai
Gulf of To

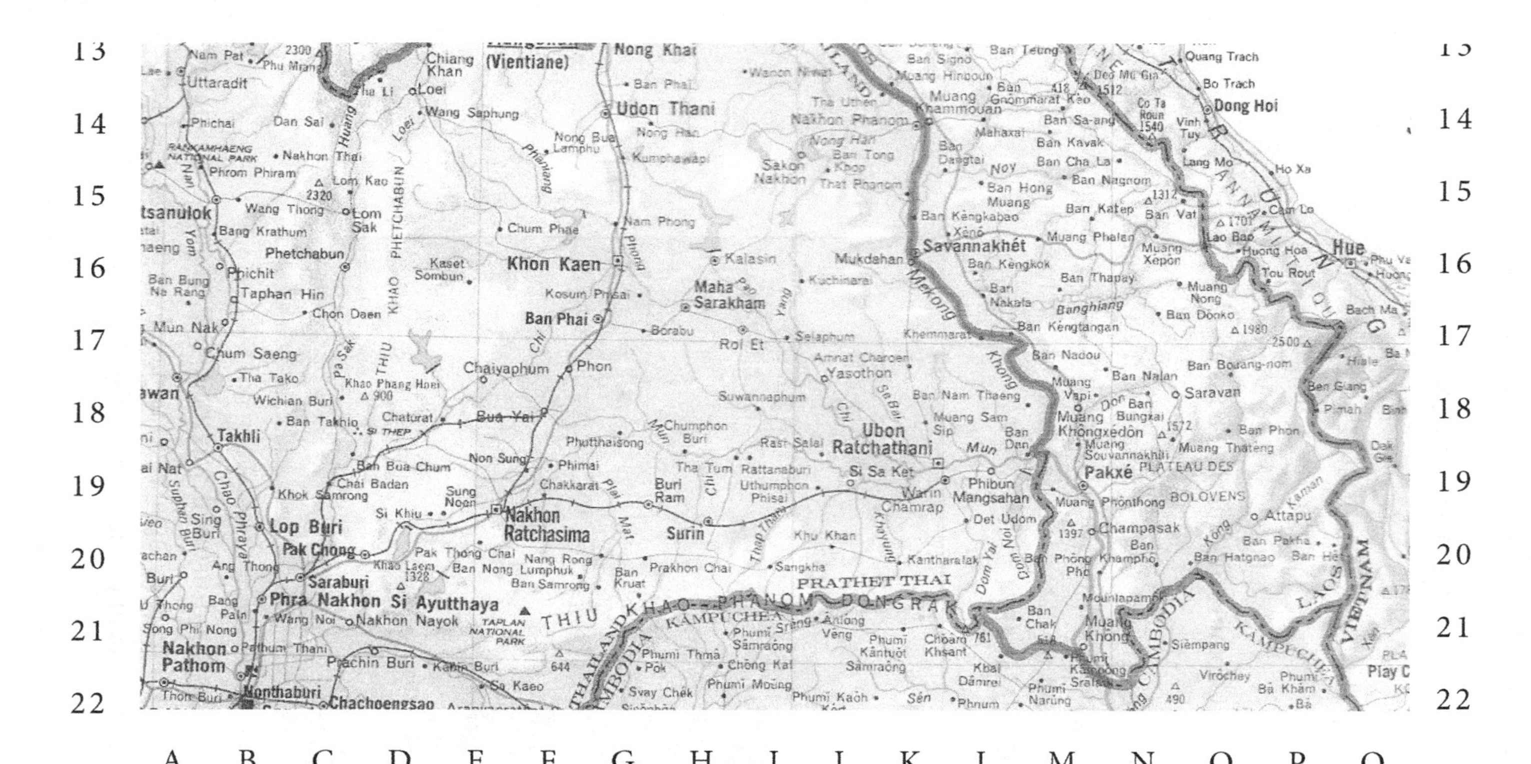

The small land-locked country of Laos shared borders with six other countries.

THE COUNTRY OF LAOS

from Reader's Digest ATLAS of the WORLD 1990, page 109

DEDICATION

This book is dedicated to the memory of
Gary James Connolly
Captain USMCR, Vietnam
Captain Air America, Laos and Vietnam

Air America rotary-wing Captain Gary Connolly celebrated his last day flying for Air America by flying his last leg naked, wearing only his captain's frame hat, sunglasses and his Nomex flight gloves. That day was early May 1974.

Photo from Connolly's archives – early May, 1974.

That was his good-natured way of saying goodbye to South East Asia after six full years flying in the Vietnam War theater. He liked to think of himself as an old China hand.

Gary Connolly flew helicopters for Air America for almost five years. He was one of the first of the new-hires when Air America started hiring helicopter pilots in late 1969. He immediately made captain, flying the H-34 single pilot around Laos.

Born in October 1942 to parents of Swedish descent, he had recruiting-poster good looks. Tall with wavy blonde hair and green eyes, he grew up near Sacramento, California, attended Sierra College, where he earned an associates degree in philosophy and became a certified SCUBA instructor.

Marine Aviation Cadet Gary James Connolly

At Sierra College he was recruited to attend U.S. Navy Flight School as a Marine Corps Aviation Cadet (MARCAD). Flying the H-34D for 13 months in Vietnam, he had many exciting and harrowing adventures. He earned 26 Air Medals and several other medals, including a Navy Commendation Medal. After Vietnam, he extended his tour and spent six months at Marine Corps Air Station Iwakuni, Japan, where he worked to increase his fixed-wing flight hours.

Sadly, his dream to become a commercial jet pilot was never to be. When he got out of the service in early 1969, the airlines were furloughing pilots. Some of our mutual friends in our reserve squadron (HMM-769) at Naval Air Station, Alameda, had worked for and subsequently had been furloughed by three different major airlines. Even though he wanted to be an airline pilot, he was also driven to succeed financially. With Air America he saw his chance to do both.

We corresponded during his first six months at Air America. He convinced me it was much safer than Vietnam. An exchange of one-word letters completed our correspondence. Hired by Air America, I sent Gary a one-word letter: "Inbound." His one-word response was, "ETA?"

Our life flight paths were parallel. We were both from Northern California. We met in Pensacola at Navy flight school, flew and roomed together in Vietnam, and were based together at the Lighter Than Air Facility in Tustin. We both flew together at NAS Alameda for the USMC reserve squadron, and again at Air America. H-34s all the way. Senior to me, he bid his vacations and R&Rs to match mine. We travelled completely around the world twice. We drank copious amounts of whiskey, wine and beer, and shared an occasional woman. We became very close.

Gary was based at Udorn, Thailand, for the first four years of his Air America experience. On a potentially hot insert north of Pakxe, I flew with Captain Dave Ankerberg. Our engine failed while we were sitting in the landing zone. Gary flew copilot for the bird that rescued us. I will always feel I owe him one. When the Vietnam peace accords were signed in Paris, he was transferred to Saigon to support the International Control Commission, (ICC) truce keepers.

With his driving desire to be an airplane pilot, every time any slot came open for an Air America fixed-wing position, he bid for it. With his low seniority, he knew his chances of being selected were slim. He also knew that if he did not bid, his chances were zero. He never got to fly large airplanes for Air America, nor for anyone else for that matter.

While at Udorn, Gary dated Barbara Ford, the daughter of our Udorn base Vice President Dick Ford.

After leaving Air America, Gary decided he must pursue his goal to become an airline pilot. In those days, airlines completely disregarded helicopter time as flight experience. To correct that, he took a huge step back from his goal to become an airline pilot and decided to become a fixed wing crop duster so he could accrue fixed-wing hours and earn money while doing it. He did not want to instruct fledgling pilots in airplanes.

When he applied for fixed-wing crop dusting jobs, he learned that he needed to have crop dusting experience. He then decided to garner some crop dusting experience by dusting in helicopters – another big step backward. When he applied for chopper dusting jobs, he was told he lacked the needed helicopter dusting experience. He then took a third giant step backward to attend helicopter crop dusting school in California's Central Valley.

He completed crop dusting school, obtained his crop dusting license, and began to accumulate experience. He had just to get a season or two of helicopter dusting experience to qualify for a fixed-wing crop dusting job. He could then begin to accrue the coveted airplane time he needed to make himself valuable to the airlines. The cut-off age for the major airlines was 31 years. Gary was passing that age and he knew he might have to settle for flying for the charter airlines instead of the majors. I believe he would have eventually worked for, and retired from Flying Tiger Airline, which was later purchased by FedEx.

Gary went to work for Northwest Helicopters based near Napa, California, just north of San Francisco Bay. By mid-1975, he was close to his first step back up, to airplane crop dusting. He delighted in saying he would never fly helicopters again after that.

On 2 July 1975, he was crop dusting vineyards south of Napa. On his first pre-dawn takeoff of the day, the tail rotor drive shaft of his Bell H-47 failed. The entire tail rotor assembly departed the aircraft, throwing the center of gravity off, causing Gary to nose over into the grapevines. The aircraft did not burn right away, but his chicken ground crew, fearing explosions, did not come to his rescue. The aircraft burned, killing Gary. He was 32.

I know his last words were, "Oh, shit!"

Bell H-47 (WIKIPEDIA) – the type of helicopter Gary died in in 1975

His mother sued. The pre-trial discovery determined that the tail rotor drive shaft had failed because it was bogus – not manufactured by Bell Helicopter Company. The inferior shaft had been slipped into the spare parts supply system and sold as a genuine Bell part.

Worn down and distressed by the trial, Mrs. C. finally settled for $60,000 for her son's wrongful death. The lawyers took almost half of that.

Someone murdered my best friend Gary for the price of a bolt.

Gary was gone, leaving a big hole in my life which to this day has never been filled. Barbara Ford was even more devastated than I. She eventually married another commercial pilot.

PREFACE

Recalled!

Mid-May 1972
Laos, Southeast Asia, about 50 miles north of the capital, Vientiane.

The cryptic radio message, "RTB Tango Zero Eight," was at once routine and, at the same time, highly puzzling. I "rogered" the message, turned my helicopter southward towards home, and wondered why I was being recalled.

RTB meant, Return to Base.

Tango Zero Eight was CIA company code for our home base at the Royal Thai Air Force Base, Udorn, Northern Thailand (*G-14 on map*), also known as simply "Tango."

I was piloting a Sikorsky H-34 helicopter in upcountry Laos for Air America, the air arm of the CIA, during the Vietnam War, on the other side (not *for* the other side) of the Ho Chi Minh Trail. I had been working upcountry, routinely delivering rice to refugee villages near Lima Site 272 (*G-10 on map*). When I radioed in for my next day's schedule, as I did every night, I fully expected to spend the night at Vientiane and do more of the same the next day. Instead, I got the urgent message "RTB" tonight … this evening … as in very soon … as in Right Now.

The message was highly unusual because we never returned to base unless we had completed our six-day trip or had serious maintenance problems that required going to the shops. It was only my second day away. My machine was in fine shape. As aircraft commander, I was the one to decide if the helicopter was broken enough to return to Tango for repairs. I was single. Both my parents had already died. I had no immediate family to be concerned about or to be worried for me.

There were no other normal reasons to RTB.

I began to wonder what was up. My own insecurities made me assume it was something I had done to get myself in "hac" Did I screw something up – again? Who did I piss off this time? What did I do to warrant being called back so early from my trip? What the hell was going on here? Why was I being recalled?

1. Could it be my bad judgment resulting in a near-fatal-fiery crash while flying with instructor "Harry" my second day of line training up country?
2. Or our drunken antics at the company bar that same night, seriously breaking the company's strict twelve-hour bottle-to-throttle rule?
3. Could it be when I seriously insulted and pissed off the base vice president?
4. Could it be that I ruined an $85,000 engine while executing the high-hover hoist rescue of an Air Force pilot and his observer?
5. Could it be that I drank wine while on duty with the French Foreign Legion paratroopers?
6. Perhaps I should not have wrestled a senior pilot for control of his aircraft and forced him to let me out of his aircraft?
7. What about when I insulted two senior pilots?
8. Perhaps it was an accumulation of all the above.

My imagination could not begin to wander far enough to guess the real reason for my recall. I was in for a huge surprise — something that would be the beginning of the end of my Air America flying career.

As a result of this recall, I would soon start flying Bell Hueys.

A few things to know before reading this book:

- The word "Customer" generically refers to the CIA, Air America, USAID, and/or the individual CIA operatives, depending on context.
- Our H-34 helicopters were unmarked except for a large H-number on the tail, i.e. H-15. Each H-34 was called on the radio by its H (phonetically: HOTEL) number, so H-34 no. 15 would be called, "Hotel one five."
- The Bell Hueys were called by their Lao registration number or letters, i.e. Bell XW-PFH became "Papa Foxtrot Hotel" over the radio. Some also had U.S. registration numbers. I never understood the difference between, or the need for, the different registrations.
- I did not take as many pictures while with Air America as I wanted to. Not long after my arrival, the company published a directive that pilots would no longer carry cameras.
- All events related here are in chronological order, as best as I can manage them. In each event cited, I cite the date, give the helicopter number, the flight mechanic (FM), and who I was flying with as copilot or pilot, if known.
- Most of our landing strips that we worked out of were called Lao Sites, or LSs; i.e. LS-272. Some of the major places like Vientiane had "L" numbers. Vientiane was L-08. I give coordinates on the map for most locations mentioned.
- I did not discover until early 1994 that I was suffering from a bad case of PTSD from the absolute carnage and destruction I witnessed in Vietnam. This may help explain some of my acting out and perhaps what drove many of my actions. Some of my Air America experiences may have exacerbated my PTSD.
- To better understand the history of Air America and the military and political situation in Laos in the early sixties through

the mid-seventies read, the reader is directed to the Air America web site history page: https://www.utdallas.edu/library/special-collections/hac/cataam/Leeker/history/index.html

- I include every "SPECIAL" mission I flew and the dates when crew members were lost during my time there to emphasize the dangers we were exposed to daily.

1

Worry One

I was wrong. Very wrong! This incident rates highest in my "Dumb Maneuvers List" of my entire 32-year helicopter flying career.

-Author

20 July 1970
Hotel 30
Flight Mechanic not recorded
Line Training with Instructor pilot "Harry" (Harry is a pseudonym. This instructor pilot asked that I not use his real name.)

The first event that crossed my mind as a reason for being recalled was my near crash with instructor Harry on my second day of routine upcountry line training. Until then I had only done routine touch-and-go landings around the Udorn, Thailand, airport and local flight checks to see if I knew how to fly the H-34. Of course I did. I had recently flown 750 hours of serious combat in this machine in Vietnam. Subsequent to Vietnam I flew an additional 200 hours in the machine. I had nearly 1700 hours flight experience overall. I had as yet flown little in Laos, but right away felt that I had more combat experience than any of the Air America instructors.

Harry was showing me the ropes. We were flying routine resupply missions to local troops in the dam site area (*G-12 on map*), about 40 miles north of Vientiane. This quiet and uncontested area would soon be submerged by water behind the under-construction hydroelectric dam. I observed while Harry did most of the flying. As he flew, he explained company operating procedures and rules and familiarized me with the area.

We flew out to one of the most challenging landing zones, a deep cookie-cutter hole surrounded by 150-foot tall teak trees. Landing here required coming into a high, out-of-ground-effect hover and then a cautious zero-airspeed vertical decent. A pilot doing one of these landings needed more than just a little experience to land at the bottom of this hole without going splat in the bottom. Harry carefully descended and expertly landed the helicopter in the center of the cleared area. As we sat there waiting for the local troops to off-load the cargo, he said those words every pilot loves to hear, "You got it." He transferred control of the machine to me.

I put my hands and feet on the controls and prepared to make the takeoff, straight up out of the hole. We had on-loaded a few Lao soldiers who needed a ride. All set, I brought the helicopter to a low hover; all felt Okay. I continued to bring the helicopter up higher and higher, until we were hovering at 150 feet. At tree-top level, we looked out at a sea of green. At this point, it became obvious to both of us that we were too heavy. There was not enough reserve power left to accelerate and fly away. The proper procedure at that time would have been for me to hover carefully back down to the ground, land and off-load a passenger or two.

Not me. My budding disdain for these Air America instructors allowed my Marine Corps pride and my youthful arrogance to override my common sense. I decided I would display my Vietnam "stuff," and show this old former Army fart how we Marines did it in Vietnam. I would do a max-adrenaline "pump-and-dump" take off.

A pump-and-dump take off was something I had done many times in Vietnam. I pumped the collective lever up to give myself a momentary rise in elevation. I knew for certain that doing so would cause me to loose RPM. Every time I had done this in Vietnam I had had one of two aerodynamic factors in my favor.

The first aerodynamic tool I intended to use was called ground effect. When a flying machine is within a wing span of the ground or water, there is a cushion of air created by the downward thrust of the wings that makes flying more efficient. Think of a pelican skimming effortlessly over the surface of the water—he is gliding on ground effect even though he is over water.

The second tool is called translational lift. Once a helicopter starts to move forward, at about 13 knots the rotor system begins to act like an airplane wing. At that point, aerodynamic efficiency increases. Many times in Vietnam, I used ground effect to cushion a landing or ease a takeoff, and a few times I jumped my helicopter off a pinnacle or a cliff into clean air, which allowed me to easily attain translational lift before hitting the ground.

I was confident I would immediately attain either ground effect or translational lift or both. I would then be able to lower the collective, reducing the need for a higher RPM-sucking blade angle, regain my RPMs, and fly away.

I was wrong. Very wrong! This incident rates highest in the "Dumb Maneuvers List" of my entire 32-year helicopter flying career.

Quicker than Harry could react, I pumped the collective, gave some forward cyclic (stick), and began to move forward. In an instant, I realized that I had made a potentially fatal decision. I had no clear air under me to dive into to gain airspeed. I did NOT gain translational lift as anticipated. Neither were the soft, leafy treetops a hard enough surface to provide ground effect.

As we slid forward out of our hover with low RPM and no maneuvering room to regain them. We began to sink down into the sea of tall teak trees. I had a fleeting vision of us hitting the trees, rotor blades bending, chopping off the tail, flying debris and tree branches rupturing fuel cells and breaking electrical cables ... the perfect recipe for fire and explosion with us in the center of ensuing fireball.

HOLY SHIT, we're going to die!

But I had no time to reflect. I fought to avoid that fiery crash. As we plummeted into the trees below us, the rotor blades chopped foliage around us. Bits of branches and leaves engulfed us, creating a green whirlpool swallowing us, blocking off all visibility. We were along for the ride … a very short ride. The more foliage we chopped, the more RPM we lost. The more RPM we lost, the faster we were sucked into the green vortex. In the center of a green tornado, there was no place for us to go but down into fiery oblivion. Out to the front and the right side, visibility dropped to zero in the middle of the green hurricane.

Looking down to the left, I barely spotted an area where the trees looked smaller and shorter. I fought to fly the machine as best I could towards that spot. I added all the power I could, pulling in full collective pitch to cushion our inevitable crash. To hell with all RPM and power limitations. If I could only get there, we might crash only once instead of two or three times as we flailed down through the trees. Like a gigantic lawn mower, we sliced and diced our way through the trees and brush.

We did not land. We simply ceased descending. We settled gently onto a small knoll covered by elephant grass. This thumb-thick grass grows as high as 13 feet. It had obscured everything on the small piece of high ground that I had "chosen" to land upon. It had reached higher than the rotor blades until I unceremoniously mowed it off to rotor blade height. I had no idea what we had landed on … or in … but we were on the ground, intact and not crispy-fried to death in a smoking heap of burning helicopter wreckage. The hole in the trees was just slightly smaller than the helicopter's rotor blade diameter.

A green whirlwind swirled around us. The smell of freshly mowed grass filled my nostrils. All I could see in every direction was a greenish fog. The helicopter windows were obstructed by a thick, green, downy coating of freshly chopped vegetable fiber. As the debris settled, I saw pieces of brush and limbs all over the landing zone (LZ), fallout from my descent. One small tree, almost three inches in diameter, was so close to the cockpit window that I reached out and touched it. It was chopped off at rotor height. I smelled freshly mowed grass.

Shit! Double SHIT! But we were still alive.

We had no choice but to shut down and assess the damage to the aircraft. Miracles do happen, I realized, as we inspected the helicopter. I had barely avoided puncturing the belly of the aircraft (fuel tanks!) on tree stumps. We determined that the only damage to the entire helicopter was that three of the four rotor blade tips were damaged.

The outer seven inches of each H-34 rotor blade is a replaceable cap, meant to be expendable and easily replaced. I had intentionally ruined a number of them while flying for the Marine Corps in Vietnam, as I mowed my way through tall elephant grass into hillside LZs or slightly too-small LZs to rescue badly wounded Marines.

H-34 replaceable blade tip like the three I damaged in the descent.
COURTESY OF H-34 CHARLIE, SANDPOINT, IDAHO

As our main rotor system chopped its way down through the trees, it cleared a path for the tail rotor. The lighter, more delicate tail rotor system had no damage whatsoever.

It took the Filipino crew chief, Harry and me, with the help of the Lao troops, about an hour to hack away the brush and trees enveloping the helicopter and to clean the greenish, thick fibrous coating off all the windows.

For the remainder of the day, we announced our presence by the whiss-whiss-whistling of the damaged blade tips slicing the air. Harry was embarrassed when he had to face up to the humiliation of calling Tango to order three new tip caps to be sent to us at our base of operations. The other pilots heard him calling on the radio, and he got a lot of ribbing. I never learned if he ever had to explain to the chief pilot about the damaged tips. If he did, he never told me.

I am surprised that Harry did not simply return to base immediately and say I was unfit to fly for the company. I think the company needed pilots badly. Had we crashed and survived, I am sure I would have been fired. One of my Marine Corps contemporaries had recently crashed near the "Secret CIA base at Long Tieng." While trying to fly up a valley through low clouds, he hit a "cumulus-hillsidulus," crashed and burned his H-34. No one was hurt, but the company frowned upon avoidable crashes because they had to explain the loss of aircraft to the government. The loss rate had gotten pretty high, so the "stuff" came downhill and rules tightened up. He was fired.

At least I showed I had some balls. I did not whimper or beg to apologize to Harry as I probably should have done for nearly killing us. I had had so many close calls in Vietnam that this little event really did not faze me. But I did know that I had to clean up my act and not do any more Nam heroics or I would be unemployed.

Or dead.

ALCOHOL
Because no great story ever began with
"Hey, want to go out for a couple of salads?"
(Paraphrase of a plaque I saw in a shop window.)

2

Worry Two

Seriously breaking the 12-hour rule

After work, Harry and I retired to the company bar on the Vientiane airport for a few drinks to de-stress from our eventful day. FAA rules state, "No drinking for eight hours before a flight." The Air America rule was more restrictive, "No drinking for 12 hours before a flight." Harry and I had a few, and it grew late. We had an early departure the next day, and we were pushing the limit of the 12-hour rule. This rule had been emphasized several times in ground school at Udorn. The company was serious about it. When I mentioned this, Harry looked around and observed that there were no other patrons in the bar. He took a big roll of Lao Kip, the local currency, out of his pocket and gave a wad to the diminutive Lao bartender. (The largest denomination of Laotian Kip notes was 1000 Kip, equal to two dollars U.S.) Harry instructed the barkeep to lock the door.

Then we got down to some serious drinking.

A 100 Kip note. The white square is a watermark picture of the King of Laos. When you hold it up to light you can see a picture of the king. (AUTHOR'S SCRAP BOOK)

After a few more drinks, Harry began to rag on me for being a Marine, and I began to jab him back about his being an "Army puke," all in good, rivalrous fun. After a few more drinks, he looked at me and said, "Watch this, Marine asshole!" He stood up, kicked his stool away from the bar, put his forearms under the bar and lifted the bar an inch off the floor. He let it drop back down with a crash and a mild banging of the glasses and bottles behind the bar.

This was not an ordinary bar, but a slab of teak two feet wide and 25-feet long, three to four inches thick. It weighed several hundred pounds, with shelves full of glassware, bottles of liquor and sinks full of water packed closely behind it.

I couldn't ignore this challenge. I still had a thing or two to show this old army puke, even though – and especially since – I had nearly killed us earlier in the day. I had to, at least, match him or lose face. I looked around to ensure the bar was still empty of others and that the door remained locked. Then I semi-squatted, put my arms under the bar and strained. I managed to lift the bar about two inches and let it bang again to the floor.

My thump was bigger than his thump.

Not be outdone, Harry took a second try at lifting the bar. This time he got it to about three inches off the floor, letting it crash down with an even greater clanking and tinkling of glassware. A few glasses clapped together and broke. Some of the bartender's tools fell to the floor in a crescendo of silverware. The little bartender started grabbing anything and everything that was loose behind the bar, placing it on the back bar out of harm's way.

It was again my turn to lift the bar, and I knew I was in deep trouble. It had been a challenge to lift the first time. I braced my arms under the bar, placed my legs in such a position as to get maximum leverage, and strained upward on the bar. It would not budge. I was about to lose this contest. I relaxed for a second, took a deep breath and strained a bit harder. The bar began to move ever so slightly. I put a surge of mental energy into to it: *I will lift this bar or break my back trying!* The bar rose to about four inches. I let it slam to the floor with a great THUD! accompanied by an even louder crashing of glasses and bottles. By this time the bartender was dancing around frantically behind the bar, beside himself, not knowing whether to run away, call the cops, the duty officer or what. He feared his job was on the line. These crazy helicopter pilots were destroying his bar!

Harry did not try to lift the bar again. I won. We drank until late, completely disregarding the 12-hour rule. I didn't realize how seriously I had flirted with getting fired. I just figured that if I was with my captain and he was drinking too, then it was all right. Many months later, I heard a rumor that a helicopter captain and first officer had gotten drunk in the VTE bar and trashed it. It took a moment until I realized the rumor was probably about Harry and me.

You might say we raised the bar for drunken antics.

At no time during this whole episode did Harry or I ever specifically mention the near-crash at the tight LZ, nor did my pride allow me to apologize for my dumb-ass maneuver that nearly killed us.

Was this the reason for my recall? Had the Company somehow

learned the identity of the two drunken pilots who trashed the bar? Could this be the early end of my Air America career?

I continued flying towards home plate, puzzled.

3

Routine Work

Everyone assumed that we worked directly for the CIA. We wondered ourselves, but we had no solid indication and thought it better to deny the rumors. When asked, our story was that we worked on contracts for the U.S. Agency for International Development (USAID) in support of the 300,000 refugees displaced by the North Vietnamese Army when it created the Ho Chi Minh Trail. And we did – some of the time.

Al Cates, in his book *Honor Denied*, explains in great detail how it has been proven that Air America was actually owned by the government. All Air America employees were in direct employ of the U.S. government. Many of the senior pilots were retired from the military, and this extra time with Air America should have increased their government retirement benefits. The bill to correct this has been buried in Congress for decades. Those pilots who put their lives on the line daily for years have been denied their earned benefits. It seems Congress is waiting for all of the older men to die. Most have by now.

At the time I went to Udorn, I believe I was the youngest Air America pilot there. Gary and Kawalek and a handful of others hired shortly before me were all older than I. All the other pilots had been there for years

The near-crash incident with Harry took place just south of our base LS-272, Ban Xon (G-10). Almost all of our routine training and check-ride flights took place at and around this base, roughly 40 miles north of Vientiane. This work was strictly humanitarian. We supported some 300,000 Lao tribes people. Had these simple people not moved west, they would have been forced to provide slave labor for the North Vietnamese Army on the Ho Chi Minh Trail. Our work in this area was always routine resupply to local civilian villages. This area was safe from enemy action.[1]

Until it wasn't.

A typical workday around LS-272 would go something like this: Each night Operations gave us our assignment for the next day. "Report to customer at LS-135 and work as directed." Early in the morning, usually about first light, we flew to the assigned airstrip, or LZ. There would always be someone there to direct us. Most of the time it would be an American customer, and he would usually be known by a code "handle." Some handles I remember are Hambone, Greek, Clean, Hog, Kayak and Mule.

Sometimes a Lao man would hand me a note written in scrawled, barely legible English. It might say something like: "Honorable pilot, Sir. Please to take 4 men and 300 kilos of rice to (map coordinates)." This might be written on a piece of a paper bag, or typewritten neatly on white paper.

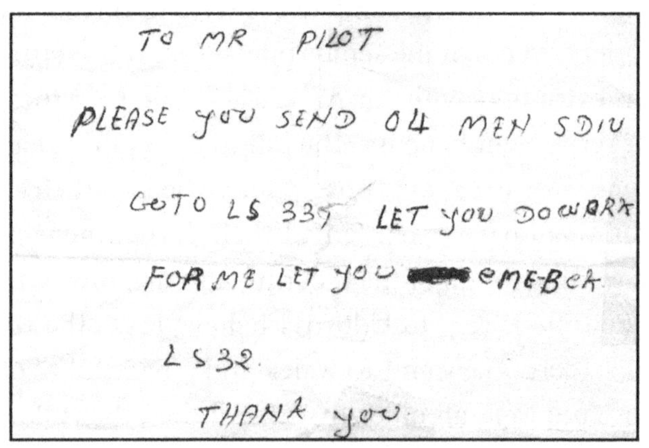

TO MR PILOT

PLEASE YOU SEND 04 MEN SDIU

GOTO LS 335 LET YOU DOWARK

FOR ME LET YOU EMEBCK

LS32.

THANK YOU.

Instructions written on a paper bag by a Lao customer.
(FROM AUTHOR'S PERSONAL SCRAPBOOK.)

Many days, I carried load after load of 100-kilo bags of rice to nearby villages. Each village had an LZ if they didn't have room to cut

1 See appendix B for an article in LOOKEAST magazine about Hill tribes people.

out a STOL (short take off and landing) strip from a nearby semi-flat piece of land.

I had a trip when, for days, I carried many hundreds of standard, ordinary milk pails out to the east to LS-266. (*H-10 on map*) I began to wonder: *What can these people possibly be doing with all these buckets?* My answer came to me on about the third day when, as I approached the village on my first load, I was nearly blinded by the sun's reflection off the roofs of all the village huts. *What the hell? Why are thatched roofs reflecting like that?* Getting closer, I saw the reason. The people were cutting the buckets down to flatten them out and then using them for roofing material. It seems that the CIA case officer for the village could not obtain roofing material, but he could order buckets.

On one occasion, just north of LS-272 (*G-10 on map*), I had to land in a large pigpen, the only space near the village big enough and flat enough to land. It held a huge sow with a dozen suckling piglets. The noise of the helicopter so excited the little piglets that they jumped around like popcorn popping for the entire five minutes I sat in the LZ. It was a delightful, fun thing to see. It was a surprise to me that little piggies could be that acrobatic.

4

Black Cloud at Udorn

My First "Special"

We almost always flew the H-34 single pilot which took a little getting used to. When we did not, it was because of missions called "specials." For these "specials" we always flew two pilots on board. Because of my low seniority, I nearly always flew copilot on these missions.

"Specials" were for doing something a little dicier than usual and entailed a higher risk of being shot at or mortared in an LZ. The incentive for specials was that they paid $50 per hour, or $50 per landing, whichever was greater. Most often they were just a flight to some remote site to insert or extract a team of infiltrators, (whom you might call spies). We might fly a recon of some area well beyond the area considered safe for normal operations.

If it were a pick up, the local troops would display a pre-arranged signal to mark their location and to confirm they were friendlies. The signals were simple but ingenious. The local troops were given two strips of bright orange cloth and were taught how to make simple figures on the ground. We would then fly over at 1500 feet, just out of effective small-arms fire range and look down on the pick-up LZ for the proper signal. The signals were either a bright orange "L," "X," "T," "I" or "V." The orange stood out well against the dark grass or brown earth. If the signal jibed with the one given us at our brief, we circled down slowly and carefully, and made an approach to the LZ.

If they did not display the proper signal, our instructions were to return to base without landing. I never had that experience.

If the mission were an insert, there was no one on the ground to make a signal. We circled down cautiously, watching for things out of the ordinary, quickly dropping our troops and departing rapidly so as to not draw any extra attention to the troops we had just dropped.

We sometimes used the same signal system for delivering food and supplies to remote friendly villages. If the landing was in a populated area, we watched the people and the village for signs. If people were about and children were playing, if pigs and chickens were visible, it was a good indication that there was nothing awry in the village. If things were real quiet, with no people and animals visible, it indicated possible trouble. We did not want to land on a helipad and be greeted by a hail of bullets or mortars. Overall the people were glad to see us coming, because it meant food, medical aid, or other supplies.

Our landing could also mean cheap and easy transportation for the Lao people. The terrain of Northern Laos is rugged. The people measured travel in days of walking. A 15-minute helicopter ride might save them three or four days of walking over rugged mountain jungle trails where tigers, bears, cobras and other hazards lurked.

4 August 1970

As a brand new first officer, I was scheduled to ride as a passenger (deadhead) with Captain Frank Bonansinga in a twin-engine Volpar, the updated, twin-turbine powered version of the venerable old twin Beech 18. As we became airborne Frank got a runaway torque indicator on the number-two engine. He made a go-around, and we returned to base to get that minor problem repaired.

Air America Volpar. FROM AIR AMERICA SITE WITH PERMISSION.

A short while later, repairs done, we took off again, only to have the problem reoccur … another runaway torque on number two. In addition, we now had an unsafe gear indication. We couldn't simply return and land because we didn't know what the problem was with the landing gear. Frank's procedures cured the unsafe gear indication by simply recycling the gear. We were able to land a second time. Frank's plane was again taken to the shops.

I abandoned Frank and hopped a C-123 cargo plane headed my way. The flight to Pakxe was without incident, which was to be the exception for the day.

(I got word in November 2015 that Frank had died. Another of the best has 'flown west' to the Great Ready Room in the Sky to await his next mission.)

A Special Too Quiet!

4 August 1970—later the same day.
H-47

FM not recorded
Copilot for Captain Dave Ankerberg

Later that same day, I flew copilot for Captain Dave Ankerberg on an insert a few miles north of Pakxe. The location was an idyllic, small, quiet village nestled on the inside curve of a large crescent-shaped field of stair-stepped rice paddies sheltered all around by tall teak trees. If you were looking for a quiet, peaceful place to escape the world, this should be high on your list as a place to investigate. (*M-17 on map*).

My job as copilot was to be watchful. I watched the instruments to make sure the engine operated properly. I watched for any movement or fire coming from the village. I peered out the left window trying to see if there was any activity in the tree line one hundred feet to my left. Everything was quiet. Too quiet. There were no people visible, no dogs running, no children playing, no water buffalo working the fields, no one planting or harvesting rice. The village seemed deserted. It was spooky as a science-fiction movie in which the space aliens have abducted all the humans. The only thing missing was newspapers blowing in the streets. But there were no newspapers and there were no streets. This could mean that the enemy occupied the village. We were on high alert as we approached the rice paddy to insert the squad of Lao troops. I had my 9mm Uzi submachine gun on my lap, loaded and cocked, my thumb on the safety toggle. Ready.

A few seconds after we settled onto the ground, the engine of H-47 made a loud cough. A big gust of dense black smoke jetted from the left-side exhaust stacks and flooded the cockpit. The engine started chugging, winding down. Simultaneously we both instantly surveyed the instrument panel and saw that almost all of of the engine gauges had pegged to one extreme or the other, and most of the red and yellow engine warning lights had illuminated. The helicopter engine had puked itself.

Dave executed an emergency shutdown. The pilots in our backup helicopter, orbiting overhead, had observed the puff of smoke shooting out of our engine and, without waiting our call for help, immediately flew down and alighted behind us. We made no delay in exiting the aircraft, running back to them and hopping aboard, abandoning our helicopter in the rice paddy. My good friend Gary was the copilot of the rescue helicopter. I remember passing a note up to him, "Is this any way to run an airline?"

A Wright 1820A 9 cylinder radial engine. (WIKIPEDIA)

When it comes to dangerous missions this one was a non-event. There was no enemy activity, and the overall situation was routine. We also knew that at any time, one of these missions could turn into a "Shit Sandwich." Occasionally one did.

On this mission, the 9-cylinder Wright 1820-84A Cyclone engine had only 13 hours on it since overhaul. The brand new impeller seal failed. Maintenance crews retrieved the disabled helicopter the next day, unmolested by the the enemy, an exception to the norm.

I will always feel obligated to Gary for the rescue, and will always lament that I was not there that awful day when he crashed south of Napa so I could have returned the favor. I would have risked a fiery explosion to pull him out.

Later that same day, I was instructed to deadhead back to

Udorn. My guess is that I was sent to Pakxe solely to be copilot for Ankerberg. Once the special mission was compromised, Dave no longer needed a copilot.

I found a C-123 going my way, but it developed a mechanical glitch and never even made it out to the Pakxe runway. My fourth downed aircraft for the day.

On the ramp sat a U.S. Army Huey, winding up, going to Udorn. I hopped aboard, thinking that, at last, I was out of danger. About half way to Udorn, the Army pilot said that he was going to have to make an emergency landing because the tail rotor drive shaft was malfunctioning. There were vibrations and smoke coming from one of the tail rotor drive shaft hangar bearings. A tail rotor failure in a Huey could get real interesting real fast.

We landed in the middle of rural Thailand, about halfway between Pakxe and Udorn (*J-16 on map*). It was getting dark. I had no idea where we were and no idea of what kind of reception we might receive way out there in most rural Thailand. From my Vietnam experience, I was paranoid about the rural people, whether they were enemy sympathizers or not. As we prepared to land, I removed my first officer's bars from the epaulets on my shirt and my Air America wings from my chest. These were token attempts to disguise my identity. I was a huge, fair-skinned *farong* (gringo) riding in a U.S. Army helicopter. If there were enemy here, we would be captured anyway, but I felt I had to try to do something – anything – to make myself less conspicuous. I learned later that the rural Thais were anti-communist and we would have been in no danger had we been forced to spend that night in a remote Thai village.

The Huey crew chief inspected the shaft. He and the pilot decided that whatever the problem was, it was not serious enough to override the get-home-itis of all of us on the Huey. We proceeded on into Udorn without further incident.

One major and four minor incidents in one day, any one of which could have turned into a disaster.

What a day.

5

First STO

Stalking Susan

Mid-August 1970,
Right after the engine failure in the LZ.

One of my buddies asked me, "Bill, have you had an STO yet?" I said, "I don't know. What's an STO?" He explained that every pilot got a week off every month. Scheduled Time Off. I had signed on with Air America expecting to get the advertised 30 days annual leave, but somehow the information about STOs slipped past me.

Wow! An R&R every month. Fantastic!

I went directly to office of the master scheduler. He acted a bit surprised, and said, "Oh, somehow you slipped through the cracks. You have a week off – starting tomorrow." Excited, I immediately told my friends of my good fortune. Dick Koeppe said he was off the same days and suggested we go to Bangkok together. We packed promptly and caught the overnight train south. Arriving the next morning, we took a taxi directly to the Dusit Thani Hotel and booked rooms.

Many years later, I was sitting in a piano bar, having a few drinks and listening to the music of Burt Bacharach. The song "The Look of Love," combined with a few glasses of wine, grabbed me by my old memories and threw me into a state of remembrance and reverie.

I remembered the night I took Susan to dinner at the revolving restaurant atop the Dusit. After a fine dinner, we danced almost every dance, as the band played nothing but Burt Bacharach songs all evening. I was amazed. I never knew he composed all those beautiful songs that I loved so much then – and still love today.

I had met Susan a few hours earlier in the lower lobby of the Dusit hotel. By today's standards I stalked her. I first saw her sitting in the hotel restaurant, eating breakfast with an older couple. Her parents? I immediately took her for a Pan American Airlines stewardess using her travel benefits to take her parents on a world tour. I walked restlessly around the shopping area adjacent to the restaurant several times, passing by and observing the trio at the table, wishing I could somehow meet that beautiful, well-dressed, perky, petite, bright-eyed, short-haired blonde. I observed that she paid for their meal, another clue that she was treating. I walked past the restaurant several more times, trying to figure out how to approach her. One does not simply walk into a restaurant and invite himself to join a family.

I did another circuit of the shops, cruising back past the restaurant. They were gone. Drat! She could have been the love of my life. Dejected and angry at myself for not coming up with some clever way to meet her, I turned into one of the shops bordering the restaurant and almost ran into the perky blonde and her parents as they exited the shop. I mumbled something extremely smooth and debonair like, "Uhmm, ahh, how's the shopping?" She seemed to understand that I wanted to meet her, and since she was traveling solo with her parents, she wanted company, too.

They invited me to join them right away on a canal tour of the city. I accepted instantly, dumping my friend Dick. After all, he and I had made no specific plans. We were in the capitol city to meet chicks. I had accomplished my mission. He was on his own, abandoned like last Friday night's empty beer bottles.

Later the four of us had dinner at the Thai restaurant off the lobby. That is when I learned her last name, as her father repeated it for the maître de. I burned that name into my memory forever.

We spent a great deal of time together over the next week and began to develop a lasting relationship. When I let it slip that it was my birthday. Susan and her mother spent hours finding a Thai bakery that could craft a custom birthday cake for me. We all shared it beside the hotel pool.

I was smitten. We corresponded while I worked in Thailand. I visited her in Gainesville, Florida, a couple of times where she was earning her master's degree in mathematics at the University of Florida, Gainesville, while concurrently being a fly-girl. During one of my visits, we spent her entire Easter vacation touring the east coast of Florida. We discussed living together. I felt we were meant to be a couple.

For years I often wondered whatever happened to her. Did she have a happy life? Did she marry? Have children? Grandchildren? An act of God kept us apart. A story for another time.

(This in no way detracts from my most happy current marriage.)

Follow up:

Susan's sister, JoAnn, married an engineer with an unusual surname. Mid-2015, I googled that name and found a realtor in New Jersey with the same unique last name. The picture on the realtor's website was JoAnn, Susan's sister. I sent a copy of my first book to JoAnn to forward to Susan. We reconnected by email. She tells me that she finished her masters' degree in math at UFG and worked for IBM for 35 years until retirement. She had been married twice, widowed twice, and never had children. She lives in Naples, Florida, snow-birding to New Hampshire in the summers to avoid the Florida heat. She raises, trains, and shows Irish Setters. She has had a life.

I was to have 27 more STOs before departing Air America, each one another adventure in itself.

I learned from Susan that the Pan American Airline flight

crews laid over not at the Dusit Thani Hotel where I met her, but at the luxurious Siam Intercontinental Hotel about a half-mile away. Pan Am had bought the entire hotel chain and was vertically integrating its business. On many subsequent visits I always stayed at the Siam. As air charter crew captains, we were automatically given a 50 percent discount; sometimes a 75 percent discount. I usually stayed at Hiltons, Sheratons or other first class hotels. This was great. I felt like royalty. I never returned to the Dusit.

6

Worry Three

I Insult the Base Vice President

There I was, flinging my wings towards Tango, trying to figure out which one of my indiscretions had caught up with me. Why was I singled out to return to Tango? Then I remembered another reason why the Company might be upset with me. On top of other small things I may have done, this may have become the straw that broke the camel's back.

One of the bennies of working for Air America was that we were allowed to get ID cards to shop at the Udorn U.S. Air Force Base Exchange. To get this desired ID, we had to go through a bureaucratic procedure that required us to fill out the inevitable forms, wait a while, fill out more forms, and then go to the USAF military police (MP) office to get a our pictures taken before finally receiving the coveted credential.

I did all the proper things in order, but each one of them turned into a quest. Every obstacle that came up required me to travel to the opposite end of the base. I did not yet have a vehicle. I had to bum rides, take taxis, or catch the non-air-conditioned base bus, which traveled slow motion via the moon. The weather was far too hot and muggy to consider walking. Besides, I was a fledgling captain. Captains don't walk. By then I was working a regular schedule. I was gone upcountry for six days out of seven. My ID card was several weeks overdue at this point in time. All my friends were now regular BX customers. I was frustrated.

Finally, I got the word that my ID was ready. All I had to do was go around to the other side of the base and fetch it. This was after the trek where the camera had broken and also after the trip where the camera ran out of film. Off I went to the MP's office. They had film; they had a working camera. They actually took my picture and glued it to the card. But then, NO! Yet another glitch. As the airman was about to laminate my card, he noticed that my card, had somehow slipped through the system without being signed by our Air America base manager, Mr. Abadie, whom I had never met. I was angry. Another delay and I had some serious shopping I needed to do at the BX. Damn!

Exasperated, I returned to the Air America side of the base. I went to the locker room to fetch something out of my locker. My friend Dennis Kawalek stood at his locker, 20 feet away. He made the mistake of asking how my day was going. I unloaded on him and told him of my frustration and the futility of my quest to gain the elusive ID. I ended my rant with "... and to top it all off, that stupid, fucking, Goddamn Abadie hadn't signed the fucking card, so I couldn't get the sonuvabitch." Dennis flinched, held up one hand and pointed behind it, *sotto voce,* toward the only other fellow in the locker room, the tall stranger between us changing clothes. Dennis then actually ducked out of sight behind a bank of lockers. It took me about two milliseconds to realize that the stranger was "that stupid, fucking Goddamn Abadie."

Oh, shit!

I knew I had to do something quickly, so I approached Mr. Abadie, held out my hand to shake his. I said, "Mr. Abadie, if I have insulted you in any way, I would like to apologize; I didn't mean anything personal. It's just that this whole BX ID card thing has me really upset." Mr. Abadie turned away from me and walked out of the locker room, refusing to shake my hand.

I soon had a message to report to Chief Pilot Wayne Knight at 0900 sharp the next morning. Terrible thoughts crossed my mind.

I was sure I would be fired from the best helicopter flying job anyone could ever hope to have. When I entered the chief pilot's office, who was sitting beside the chief pilot but the dreaded Mr. Abadie?

I was doomed!

Much to my surprise and relief, Mr. Abadie reached out, shook my hand and apologized to me for not accepting my apology the day before. Phew! Relief! I became more careful about how I talked about people after that.

Could this insult to Mr. Abadie be the reason for my recall?

I flew on southward, more puzzled with each beat of the rotor blades.

Follow up:

Nineteen years later, in 1989, I applied to Johnson Controls to fly Hueys on Kwajalein Island in the Marshall Islands for the U.S. Army missile testing system. The same Mr. Abadie was the base manager there. I was worried that my previous encounter with him might harm my prospects of getting hired for the new job. During the application process, I did not feel comfortable asking him about the incident. He hired me, so I guess he had forgotten it.

At an Air America reunion in Las Vegas in June of 2001, I related this story to Mr. Abadie as I remembered it. He said he remembered the incident. He reported only that he had walked away from me, thinking to himself, "Yes, that Abadie really is an asshole!"

7

Background

I hired on with Air America with trepidation. I first saw an Air America Huey in 1967 in Dong Ha, Vietnam, while flying for the United States Marine Corps. When I spotted the silver and purple Huey refueling beside us, I asked my captain, "What the hell is that?" He replied, "CIA." Then I saw the small lettering on the tail of the Bell Huey, *AIR AMERICA*. I replied, "I don't want anything to do with that spy business. Too risky for me."

Less than three years later, I found myself flying for the "Company." How did that happen?

I flew H-34s for the USMC in Vietnam from late July 1966 until mid-August 1967. For details of that adventure, see my book:

The Adventures of a Helicopter Pilot
Flying the H-34 in Vietnam for the United States Marine Corps
(Available on amazon as a paperback or an ebook.)

I accrued 750 hours of combat flying, during which I hauled about 600 wounded Marines out of the field. I had seen enough carnage and destruction to last several people many lifetimes. I wanted nothing more to do with the Vietnam war.

When I returned to the United States in August of 1967, I had figured out the meaning of the word officer: "One who works in an office." I decided I did not want to be a career officer. I was a pilot; I wanted to fly full time for a living.

Another factor for my getting out of the Marines was the tandem rotor H-46 "Sea Knight" helicopter. This Boeing-Vertol product was the latest technology and it was gradually replacing our venerable H-34s in the USMC inventory. It was a sweetheart to fly, a stable platform for flying on instruments in the clouds. It had twice the power of an H-34 and could carry twice as much. It had twin-turbine reliability.

Unfortunately, it also had a series of "developmental" problems with a tendency to crash (read: come apart in the sky!) without warning. Several of these H-46s crashed during the developmental stage, killing many Marines. I had witnessed the aftermath of several that had crashed around I Corps in Vietnam where I flew until July 1967. One area, just south of the DMZ near the western foothills, became known as "Helicopter Valley" because five H-46s were shot down or crashed in one spot on an early morning insert. They burned for days. Sometimes our brief at Dong Ha was: "Go northwest about 12 miles, turn left at the burning H-46s." (For details see Marion Sturkey's *Bonnie Sue, see bibliography*)

Developmental problems are common in new aircraft. Many men have died in advancing aviation to where it is today, but I felt that I did not want to sacrifice myself to this on-going experiment.

The fellows who flew this machine later in Vietnam, and even later in Iraq and Afghanistan, love it. The H-46, in turn, has recently been phased out in favor of the newer Boeing Vertol V-22 Osprey.

H-46 Sea Knight "Phrog" (AUTHOR'S PHOTO)

V-22 Osprey (AUTHOR'S PHOTO)

I felt that to stay in the Marine Corps was a death sentence. I was drinking prodigious amounts of booze at that time, and I knew that if I returned to Vietnam, I would continue to get drunk nightly. I knew I would not survive the triple whammy of going back to the war, flying the H-46A in combat and constantly drinking to excess.

I now considered myself a professional pilot. I wanted to fly for a living. I exercised my reserve contract and left the Marine Corps in December of 1968. Unfortunately, the airline situation in the late 1960s and early 1970s sucked. With the economy in the midst of a recession, airlines were not only not hiring, they were furloughing pilots in droves. One friend in my reserve squadron had been hired, trained, and consecutively laid off by three major airlines.

Just before we got out of the service, my good friend and fellow pilot, Dennis Kawalek, whom I had known since flight school in Pensacola, hired an international employment agency near Los Angeles to send his resume to 250 airlines all over the world. If U.S. airlines were not hiring, then perhaps some foreign airlines were. I hired the same job agency. As a result of that mass mailing, I got an application for Air America. I knew that Air America was the CIA and hiring on with them meant returning to Southeast Asia. I filled out the application and mailed it anyway. Just in case.

In the spring of 1969, I took a job flying an old military H-19 in Alaska, helping with the oil lease explorations on Alaska's North Slope. In late October of 1969, I returned to my aunt's address in Napa, California. She handed me a telegram that had arrived in my absence. It was from one Mr. H.H. Dawson with the simple message:

"If interested in employment as a helicopter pilot in S.E. Asia, call me collect: xxx-xxxx."

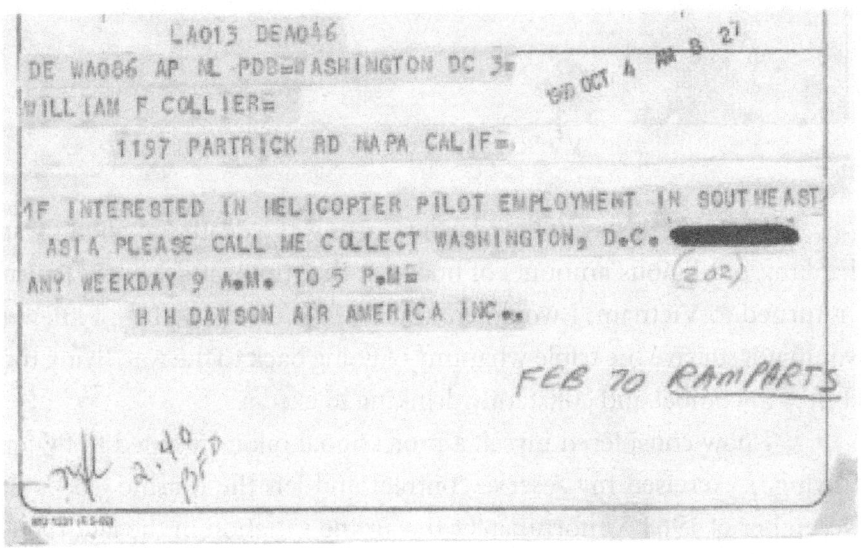

LA013 DEA046
DE WA086 AP NL PDB=WASHINGTON DC 3=
WILLIAM F COLLIER=
1197 PARTRICK RD NAPA CALIF=.

IF INTERESTED IN HELICOPTER PILOT EMPLOYMENT IN SOUTHEAST ASIA PLEASE CALL ME COLLECT WASHINGTON, D.C. (202)
ANY WEEKDAY 9 A.M. TO 5 P.M=
H H DAWSON AIR AMERICA INC..

FEB 70 RAMPARTS

The telegram from the CIA that started this adventure.

At that time, I was in love with and engaged to a young lady who lived in Morro Bay, California. I was not especially interested in returning to the war. That was about to change. After my extended absence in Alaska, my love decided that I was too absent from her life to be a good partner. I also was looking for more work as a helicopter pilot. She fired me.

I was free to move about the world.

A big factor in my deciding to go ahead and apply to Air America was that my good friend, Vietnam hootch-mate and flying buddy, Gary Connolly, had been hired six months prior. As we corresponded, he assured me that flying in Laos for Air America was much safer than flying for the USMC in Vietnam, and the pay was much better. As time passed Gary's letters became increasingly interesting.

In late spring of 1970, I re-evaluated the telegram. I called the number and talked to H.H. Dawson. At first he did not remember my name, as the telegram was several months old. I convinced him that I had, indeed, received it and that I was, indeed, interested in flying helicopters in Southeast Asia. He hired me.

I began the steps to prove my worthiness for the job. I took and passed a flight physical exam with a doctor provided by the company. I obtained letters of recommendation, sent copies of my logbooks and my honorable discharge from the U.S. Marine Corps.

I took the exhaustive airline pilot's aptitude test called the STANINE. This was a strenuous, all-day affair. I was beat by the end of the day, but I must have passed. A former Army helicopter pilot took the test with me, but I never saw him again. Who knows? Maybe the CIA only hired guys who flunked this test.

I did all the other things as requested, and in a few weeks I got my travel papers and airline ticket to Taipei, Taiwan.

Just before I left Sacramento,
the number one hit song in the U.S. was
"The Long and Winding Road"
by the Beatles

Feeling that I was on my way, an exchange of one-word letters completed my correspondence with Gary. Hired by Air America, I wrote,

Dear Gary, (early May 1970)

Inbound.

Wilco

His one-word response was:

CHAIPORN HOTEL

ชัยพร โฮเต็ล

209-211 Makang Road Udornthani Thailand Tel. 193

209-211 ถนนหมากแข้ง จังหวัดอุดรธานี ประเทศไทย โทร.193

28 May 1970

Wilco,

ETA?

Cons

(Sometimes Gary went by the nickname 'Cons.')

"Life is either a daring adventure or it is nothing at all."

-Helen Keller

8

Off on a Grand Adventure

MEANTIME, BACK IN THE REAL WORLD:

24 JUNE 1970

THE U.S. SENATE REPEALS THE 1964 GULF OF TONKIN RESOLUTION.

I was off on the grandest of adventures. Or was I?

When I got to the Sacramento Airport, I called in to the home office in Washington, DC, as instructed, for a last-minute check in. DRAT! Some glitch in my paperwork. H.H. Dawson told me to go home and wait for it to clear. I replied, "Bullshit! I am at the airport. I gave up my apartment. I have no home to go back to. I sold my car. I have my ticket in my hand, and I am getting on this airplane!" It seems that the paperwork caught up with itself, and I never heard another word about this from anyone.

Changing planes at San Francisco, I was delighted to encounter several of my buddies whom I had known through flight school and in Vietnam. When I said, "What the hell are you guys doing here?" They said, "Same as you." We picked up another two guys in Hawaii. By the time we arrived in Taiwan, eight of us new-hires were traveling together. The other seven guys were Dick Koeppe, John Ferris, Byron Ruck, Jim Depuy,

Newell, Mitchell and Rankin. Ferris, Depuy and Koeppe were contemporaries from Vietnam. I had gone through the Marine Corps aviation cadet program in Pensacola with Ferris and Depuy.

We spent a few days in Taiwan doing more processing and having company briefs about Air America operations. It all seemed silly to us, but heck, we were on the company payroll now. If they wanted us to fool around in Taipei, we could do that. At one point I was filling out a company form when my pen ran out of blue ink. I picked up a pencil and started filling out the form with pencil, when the point broke off. I borrowed a black-ink pen and finished up the job. Later, someone told me I had to sign the form. The only pen available for me to sign with had red ink. I used that to sign the form.

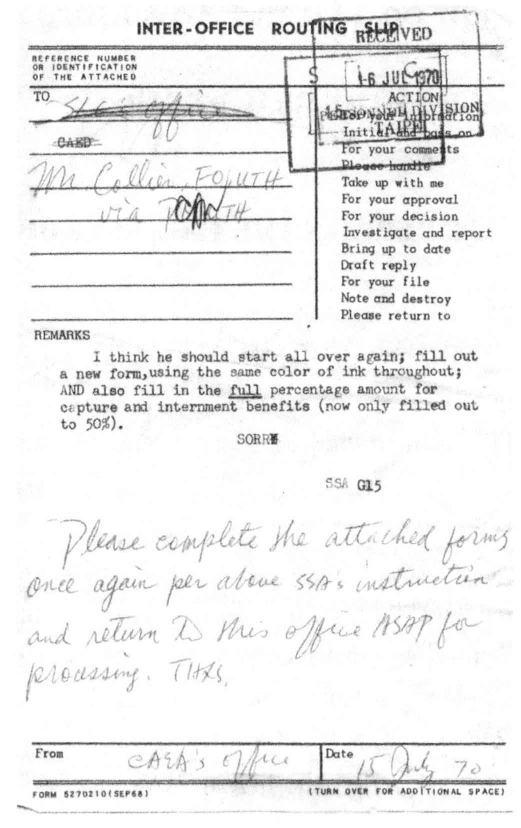

INTER-OFFICE ROUTING SLIP

RECEIVED 16 JUL 1970 ACTION PERSONNEL DIVISION TAIPEI

REFERENCE NUMBER OR IDENTIFICATION OF THE ATTACHED

TO ~~SSA's office~~
~~CAED~~
Mr. Collier, FO/UTH via [illegible]TH

For your information
Initial and pass on
For your comments
~~Please handle~~
Take up with me
For your approval
For your decision
Investigate and report
Bring up to date
Draft reply
For your file
Note and destroy
Please return to

REMARKS

I think he should start all over again; fill out a new form, using the same color of ink throughout; AND also fill in the full percentage amount for capture and internment benefits (now only filled out to 50%).

SORR[illegible]

SSA G15

Please complete the attached forms once again per above SSA's instruction and return to this office ASAP for processing. THXS.

From CAEA's Office
Date 15 July 70

FORM 5270210(SEP68)
(TURN OVER FOR ADDITIONAL SPACE)

The form that informed me that I had filled out a form too informally.

(FROM AUTHOR'S SCRAP BOOK.)

A few weeks later, in Udorn, I received that form back in company mail pouch with a request that I use only one color of ink to fill out a new form. The amusing thing about that notice was that the form was partially typewritten, partially hand-written and finished off with two different colors of ink, and had things scribbled out and corrected in pencil.

It was more of a mess than my original form.

As part of yet another Company physical, we were given small plastic cups and instructed to return them with a stool sample. This was a strange and different experience, but we figured out how to use a small stick provided to get a small sample into the tiny cup without making a stinky mess.

I flew on to Bangkok International and caught a day train to Udorn.

9

Settling into Udorn, Thailand

After my arrival in Thailand, I took the all-day train from Bangkok to Udorn. I arrived in time to spend the night in the pilots' transient housing in the company compound. The next morning, I walked to the company restaurant for breakfast, and who should be sitting at the next table but my good buddy, Gary? We had a great reunion. He took me under his wing and began to show me around.

I followed Gary's lead and moved into the Chai Porn Hotel. (The word Porn in Thai has no resemblance to its meaning in English.) We junior Air America pilots took over the entire eighth floor. It was open door policy, dormitory style, whenever you were home, you left your door open and your buddies knew you were there. When I arrived it was Connolly, Koeppe, Ferris, De Puy, Ruck, Steele, one or two others and me. Hotel management rarely put anyone else up there with us. We always had someone to drink with, which was our major form of recreation. Rent was just over a hundred dollars a month, which included linen and housekeeping.

There was a nice little restaurant downstairs that served a fair meal, and we could buy water buffalo burgers and fries for a few dollars. I occasionally went down for breakfast. My favorite meal was a stack of pancakes with a side of fruit, usually pineapple, papaya, and banana. I may have grossed out the locals when I began eating my meal by putting the fruit on top of the pancakes and then pouring syrup all over the whole thing, but I liked it that way.

The company ran a VW minivan around town to transport people to work, so all I had to do on work mornings was get myself down to the lobby, and get my sleepy body into the van. The company also had a nice little American-style restaurant at the base, so I almost always ate there before going to fly. Prices were low and the food was adequate. We had an officers' club and a swimming pool, handball courts and tennis courts. We also had a movie theater with a movie every night. The USAF base also had a Base Exchange, an officers' club and two additional movie theaters. For the married guys with children the company maintained a K-12 school on the premises. Some of their wives taught in the school.

AIR AMERICA LOG ★ 美國航空公司雜誌

Udorn's Powerhouse where diesel-driven generators supply all the electricity for the Base. Bridge in foreground is over the Mad River.

Udorn's Water Plant incorporates modern water purification system.

AAM's Traffic ... pp. 2-3). Passe... Warehouse in ba...

Spread at UDN's Club Rendez-vous. Center rear is Mr. S. Y. Chang, Chinese Headwaiter in charge of all Club dining room employees. At his left is Thai waitress Miss Lumyai Boonyen; at his right is Thai waitress Miss Sommrit Komchompoo.

Aerial view of main complex, AAM/UDN. Shown are: (1) Admin/Ops. ... Rendez-vous Dining Room; (4) Club Office; (5) Club Pool; (6) Club Pool-Side S... Main Gate; (9) School House; (10) BM's House; (11) Utility Bldg.; (12) Old ... Bar; (14) Hangar Boys; (15 & 16) Supply Warehouses.

Udorn's Engine Build-Up Shop. Row of Wright R-1820-84 powerplants for Sikorsky UH-34 helicopters can be seen at right.

Technical Services – Administration & Training Building.

Air Ame...

"NAME THE CHALLENGE—WE CAN MEET IT"

Our Air America home at the Royal Thai Air Force Base, Udorn, Thailand

AIR AMERICA LOG ★

UDORN: AIR AMERICA'S BIG BURGEONING BASE

by: *C. L. Lane, AABM/UDN*

The mission of Air America, Udorn, is to fulfill contractual obligations to various agencies of the U.S. Government. This mission covers a broad scope of activities in Southeast Asia — such as airlifting passengers and cargo, aircraft maintenance, and many related projects. That Udorn has met all its obligations competently is best demonstrated by the Base's rapid expansion and its swift personnel build-up.

Udorn's flying fleet is composed of a variety of rotary- and fixed-wing aircraft; helicopters predominate however. An average monthly fleet count would be approximately 29 choppers — both piston engine and turbine powered — versus only two or three fixed-wing planes. The aircraft were selected to best meet the basic requirements of airlifting passengers and cargo into sometimes very difficult areas.

Because Air America aircraft operate over very rugged — and sometimes insecure — terrain which is virtually inaccessible to ground vehicles, the highest overhaul standards are mandatory. Udorn's modern maintenance facilities, coupled with a cadre of highly trained, dedicated technicians, assure that operational hazards are kept to an absolute minimum. This maintenance force is kept current on latest developments through vigorous and continuing training programs.

In addition to routine maintenance, AAM's Udorn shops have performed remarkable mechanical feats in rebuilding badly bent flying machines.

All departments of the Udorn Base — Flying (Fixed-Wing & Rotary-Wing), Operations, Traffic, GMD, GTD, Accounting, Personnel, Security, Supply, Medical, Fire Brigade and Club Rendez-vous, take pride in their individual accomplishments. But they fully realize that teamwork and cooperation among all departments is essential to a successful and efficient overall organization. That is their goal.

More of our home base at Udorn.

"PROFESSIONALISM THROUGHOUT"

10

Training

We had a two-day ground school on the H-34. For those of us who had flown it in Vietnam, it was a breeze. Then we began to learn the Air America procedures and routes.

11 September 1970
H-73
FM not recorded
Training with senior instructor Captain Jerry McEntee

My log book entry says, "One full auto; engine quit." I have absolutely no recollection of this, my only-ever full-on engine failure. I have always thought that I had gone my entire career without an overt engine failure in the air. It must have been so routine that it does not stick out in my mind. I had not yet made captain, so I must have felt comfortable with my aircraft commander, senior instructor McEntee, who landed without incident.

Later, I was flying with Jim Rausch. We had just departed Udorn and were only about ten miles from the Udorn airport when we heard and felt a loud BOOM! The aircraft continued to fly normally, no instant engine failure here, and all the instruments showed normal readings. Rausch decided to make a precautionary landing anyway. I was in agreement. With these older machines and their recip engines, we were always wary of engine failure.

The U.S. Air Force H-43 Pedro search and rescue helicopter

from Udorn showed up almost immediately to see if he could assist. By then another Air America H-34 had dropped by and offered us a ride home, so we went back to home base with one of our own. The H-43 Syncropter was a weird machine, and we were a bit wary of riding in it.

US Air Force "Pedro" H-43 rescue Synchropter out of Udorn
(WIKIPEDIA)

A recovery crew inspected the helicopter and flew it back to base right away. There was nothing wrong with it. We found out later that the loud boom we heard was because we had flown directly over a Thai Army hand-grenade training field. The loud boom we felt and heard was from a hand grenade exploding several hundred feet beneath us.

17-18 September 1970
H-73

Flew 3 specials near Louang Phrabang (E-9) as copilot for Captain Frank Stergar

123 September 1970
H-57

Proficiency checkride at Udorn with instructor Captain Danny Carson.

11

I Check Out as Captain

24 September 1970
H-39
Instructor Captain Danny Carson

I flew a route check with instructor Danny Carson in the LS-20A (F-9) area. Exactly three months after getting hired, I was upgraded. I was now a certified air charter captain. This was the second most exciting day in my aviation career.

The most exciting was the day I landed my North American T-28C Trojan fighter-bomber on the aircraft carrier USS *Lexington* in the Gulf of Mexico. That day, 7 September 1965, I earned the coveted golden wings of a U.S. Naval Aviator. Thirteen weeks later, after I finished helicopter training, my girlfriend, Marci Floyd, pinned those gold wings upon my chest.

Fewer than five years after getting my Navy wings, I was an air charter captain. When I joined the military, I thought that I might make the Marine Corps a career. I also thought I might get out of the service and become a civilian airline pilot. Being a helicopter captain with Air America was certainly more interesting and challenging than flying long overseas legs in a Boeing 707, and it paid as well as what airline captains earned at the time.

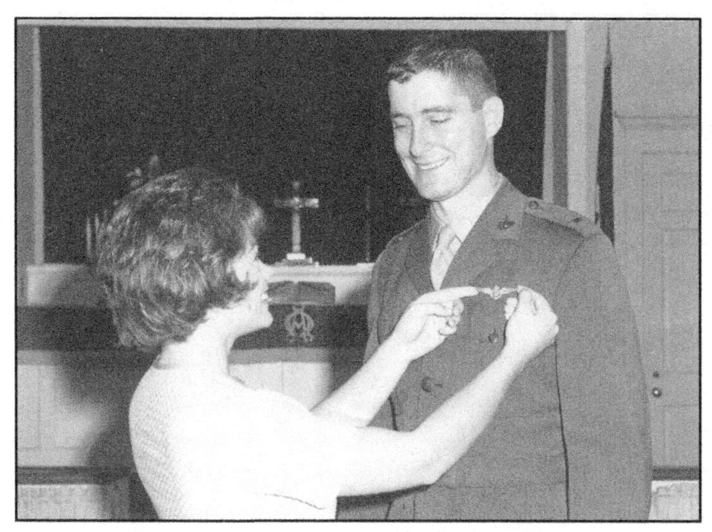

Marci Floyd pins my U.S. Navy wings of gold upon my chest.
15 December 1965. (AUTHOR'S SCRAP BOOK.)

My brand new Air America captain's wings.
Photo by Gary Connolly

3 and 4 October 1970
H-32
Copilot for Captain Hal Miller. We flew one Special each day near PS-22 (O-19 on map).

20 October 1970
H-32
FM Freedman,
L-39 (K-15 on map)

This date I flew my first solo Pilot In Command mission. It was weird at first to fly without a copilot. I was so used to having someone there, to help me see air traffic and other hazards. As it was, we rarely flew anyplace where there was lots of traffic. In those places where there was traffic, we flew along certain routes, and we almost always knew where everybody else was in our area. We also had routine quarter-hour position reports that we listened to faithfully. If another pilot reported being in the same area, we talked to him and coordinated our activities.

The basic rule of helicopter flying near airports is, "Always, avoid the flow of fixed-wing aircraft." Whenever we approached an airport, we did not make normal flight pattern with entering a downwind, turning base onto a final approach. We approached the airport at 90 degrees to the normal duty runway and slipped into the airport under the normal flight approach pattern altitude.

Position reports

One of the procedures used by Air America to keep track of all the aircraft was a simple position report. Each aircraft checked in every 15 minutes. A sample is "Oscar Mike, (operations manager) Hotel 54, three miles east of LS-272." "Roger H-54," would be the reply. Each aircraft had an eight-day clock with a timer button. After I made a report, I punched the timer back to zero, and it started ticking off the minutes again.

If someone failed to report in, a search would be initiated for the non-reporting aircraft. At first Oscar Mike would make radio calls to try and raise the non-reporting aircraft. If that did not get an answer, then OM would start asking other aircraft in the area if they had seen the missing aircraft. It was only after an aircraft was missing for a couple of hours that an actual search would start. This was rare, as all of us were responsible about making our reports.

The main advantage of our quarter-hourly reports was that when someone did appear to be missing, the last report from an aircraft gave a starting place for a search. Usually, even when someone crashed, the pilot would survive. We all carried survival radios and could communicate our location after crashing.

12

Second STO

Chang Mai, Thailand

September 1970

My second STO came up. I had been to the local bank and cashed a check for several hundred dollars. At that time, the largest bill in Thai money was a pink-ish 100 baht, which had a value of $5 U.S. I had quite a roll of bills in my pocket. I was careful to reach into my pocket and peel off just a note or two at a time and not reveal my huge wad. I was going solo, and I had no specific plans for this week off.

There were no 500 or 1,000 baht notes while I was in Thailand.(WIKIPEDIA)

As I bounded down the stairs to the hotel lobby to depart for places unknown, I spied a tall, attractive American brunette at the hotel counter was struggling to communicate with the Thai desk clerk. I volunteered to help. She was a Red Cross "Donut Dolly[2]," just arrived from Vietnam, travelling home to the U.S. After her tour in Vietnam, she had stopped over in Thailand to do the touristy bit. We talked for a while. I invited her to lunch. Over lunch, we hit it off and agreed to visit Chiang Mai, Thailand, *(east of A-10 on map)* together. Off we went for a delightful week together in northwestern Thailand. I remember we visited the silver jewelry factories, the Celadon pottery factory and the factory where the Thai women made the little paper parasols that your bartender puts in fancy drinks.

From September 19 to October 3, 1970
the number-one song in the USA was
"Ain't No Mountain High Enough."
by Diana Ross.

2 The Donut Dollies were Red Cross volunteers who served the troops by serving drinks and meals and cheering up the troops. For an informative and interesting article about the Red Cross Donut Dollies, see VIETNAM Magazine, April, 2012, p. 50-55. To view a video about these delightful young ladies, see: https://en.wikipedia.org/wiki/A_Touch_of_Home:_The_Vietnam_War 27s_Red_Cross_Girls.

13

Dangerous uniforms

Gary and I had flown the H-34 in Vietnam. We knew how flammable it was, and we were not happy with the Air America uniforms made of polyester fabric. We knew that in a fire the polyester would burn and melt into our skin. There is nothing like wearing napalm close to your skin. We talked to the chief pilot. He permitted us to wear our old military flight NOMEX suits over our company uniforms so long as we were away from home base. We agreed with that, but started slipping into our green military NOMEX coveralls just before we climbed into our aircraft to depart Udorn.

About a year later, the company decided we must have NOMEX uniforms immediately. To expedite the process, all pilots received yards of NOMEX cloth and was given money to hire a local tailor to have uniforms made ASAP.

Gary and I were then enthralled by the "Peacock Revolution" taking place in men's fashions at the time, and we decided, much against company directives, to have shirts made with pointy collars, and had bell bottoms built into our uniform trousers. We only did mild alterations so as to not to be too noticeable, but we expressed ourselves a bit.

We also worked with Abdul, our Indian tailor, to have clothes custom made. I ordered barrel-sleeved shirts made out of purple striped taffeta-like material. I had Abdul custom-build me a pinstriped, double-breasted suit with gold buttons—and a pink shirt to match. We got a lot of our ideas from old copies of GQ Magazine, the men's fashion magazine we found in Abdul's shop. We were pretty fashionable for a while. This was great fun!

14

Savannakhet (SKT)

October 1970

Another of our bases was Savannakhet, (*L-16 on map*) Laos. A day's work at Savannakhet included a lot of waiting, quite the contrary to most of our other bases. During the usual tour of six days there, we might fly only one hour. The rest of the time we lazed around the hostel, trying to keep busy with reading, sleeping, eating, or whatever. Sometimes we could get enough people together for a card game, but usually not. I spent a lot of time writing letters and reading up on investments.

Gecko Racing

It was here that some of the guys invented "gecko racing." These little lizard-like creatures abound in Southeast Asia. Their presence is tolerated and even encouraged because they are voracious insect eaters. The locals considered them to be good luck. Dick Koeppe and some of the others created a game of catching the little amphibians and putting them into a bowl of ice to put them into a state of hibernation. After the guys had caught a few, they made sure the geckos were all sufficiently chilled and then laid them out in a row on the table. Then the guys bet on which gecko would first come back to life and scramble away.

The Ant Vampire

I sat in one of the big chairs one evening after dinner, reading. A bit of motion caught the corner of my eye. I looked at the wall, and spied a trail of ants running up and down the wall. *Nothing unusual about that.* A bit later, motion caught my eye again. I glanced sideways, and I saw an ant fall to the floor. Looking down, I noticed that on the floor at the bottom of the trail of ants lay quite a collection of dead ants. *What was killing them?* As I watched, another ant fell onto the pile. I looked up the wall. Beside the trail sat another larger insect. About twice the size of the average ant, the strange creature looked like a cross between an ant and a spider, with black appendages, and a bright red body. This "ant vampire" lurked beside the ant trail, snatching ants from their trail. It sucked the juice out of their tiny bodies and cast the dried out little carcasses to the floor. I was glad to know it was so tiny and not a predator of humans.

Friendly Fire

One day the customer asked me to deliver cargo to a Lao military compound a few clicks (kilometers) out. I had no qualms about going out single ship, single pilot. This was a milk run. If I ran into trouble, the second ship was available to come bail me out.

After dropping off the cargo at the friendly LZ, I took off and made the usual precautionary, spiral climb out, staying as close over the compound as I could in case of engine trouble. At about 500 feet above the ground, a burst of AK rounds passed through my airspace close enough that I heard the rounds passing and saw tracers streaking nearby in the daylight sky. Considering my position and the time of day, these could only have been "friendly" rounds. Some idiot local trooper below decided to take a shot at me, just for the fun of it.

I reported this incident to the customer, and his reaction was a shrug of the shoulders with a "so what" attitude. He offered to sign me off for a special, but I declined. I did not care about the money. My concern was about safety – keeping our aircraft and crews safe from "friendly fire." When I returned to Tango, I exercised the only real option I had. I wrote up a Captain's Trip Report to inform management about this incident. I never heard any feedback.

We pilots quickly developed the belief that the customers considered us, our machines and our crews to be expendable. They would sacrifice us without remorse to accomplish their goals. Always wary of their instructions, we proceeded with caution whenever they sent us out on special missions. To get around them and to placate them, we always agreed to do whatever they wanted, but then we would go out and evaluate the mission from on high. Rarely did we fail to accomplish their goals, but we did it our way.

Three great advantages of flying for Air America were that we almost never had to fly into hot LZs, we almost never flew at night and we rarely flew emergency medevac missions. I did fly one night emergency medevac later.

15

MEANTIME, BACK IN THE REAL WORLD:

12 NOVEMBER 1970

THE MILITARY TRIAL OF LIEUTENANT WILLIAM CALLEY BEGAN AT FORT BENNING, GEORGIA, FOR THE MASSACRE OF CIVILIANS AT MY LAI, VIETNAM.

Strawberry Ice Cream

Pitsanoulk, Thailand
H-52
FM Tolentino
T-603 (Southwest of Bangkok).

Oscar Mike dispatched me to Pitsanoulk in southern Thailand near the top of the Malay Peninsula. This was the only time I did not go north to work in Laos. At this secret CIA base, I was to familiarize Thai mercenary recruits with helicopters. It was simple, routine stuff: take the troops for rides around the traffic pattern and help their trainers demonstrate rappelling. No combat flying here and minimal risk.

We had a little event en route that might have been a disaster, but we did not know that until hours later. About an hour into the flight south I smelled something that smelled strongly like scalloped potatoes baking. I checked all my instruments, but saw nothing amiss. I asked Tolentino if he was eating something. The Filipinos sometimes ate baluts, fermented fertilized eggs. They stunk, but were said to be delicious. I never tried them. He said no, he was not eating his lunch, but he noticed the smell, too. We could not figure it out, so we ignored it. An hour after we landed, Tolentino inspected the helicopter and discovered that we had had a run-away charge on our NiCad battery. Way overcharged, it could have exploded in flight. Had it done so, Tolentino might have been seriously injured or killed as he sat only inches away from that potential bomb with only a thin sheet metal wall between him and the battery. It might have resulted in a fire, and things would have quickly become most interesting from there. Tolentino ordered a new battery and installed it in short order. Those guys were good.

Our crew chiefs flew all day with us, supervising the loading and unloading of cargo and passengers. They put their lives in the pilots' hands hour after hour, day after day, landing after landing. Without them, the machines would have deteriorated to unflyable junk in less than a week. After working all day in the air, they often worked all night to maintain the helicopters. Sometimes they had to wash blood or puke off the cabin floor. One time a Lao villager puked all over the floor of my H-34. As I deplaned, I saw what looked like a large night crawler-type worm in the puddle of her puke.

These fellows changed out engines overnight, replaced gearboxes, rotor blades and many other components – in the field. They filled about 100 Zerk fittings with grease every night. (There must be 50 Zerk fittings on the H-34 rotor head alone.) They did all the refueling when we stopped to refuel. These were the most dedicated crews that could possibly have existed, anywhere, anytime, any war. They never got enough thanks and credit for their hard work.

Follow up:

About the year 2010, I made contact with former Filipino FM Ranny Lacson. I was trying to find FM Ernie Cortez so I could send him a copy of the DVD about the "Rescue of Raven 1-1." (*See Chapter 40*). We communicated by email, and then he and his wife and two grown children drove over from Seattle to Sandpoint, Idaho where I live, to visit. We went out to a nice lunch and caught up. One thing that Ranny told me made a huge impact. He told me that a lot of the Air America pilots were quite racist and abusive to the Filipino FMs. He told me that he and all the other Filipino FMs liked working for Gary and me because we treated them with respect and as equals.

My question is: Why would we not? Especially when our lives depended on their work. Sure, they rode with us all day long on every trip, so they looked out for their own lives too, of course. But they could have made life miserable for us if they had wanted to. They were nice folks, and we knew they were our equals. I have nothing but fond memories of working with those great fellows.

Two senior and highly experienced Green Beret sergeants ran the secret base at Pitsanoulk. They had their own well-stocked officers' club, just for the two of them. They had a story or two to tell. As in most cases, they couldn't come right out and tell me about their experiences. It took a couple of days hanging out and drinking with them in their two-man NCO club before they began to loosen up. As we began to develop a mutual trust and affection, the serious stories started to flow. Scratch any Vietnam vet, or any combat vet whatsoever, and all you will find is a hard layer of protective substance. I'm talking here about any vet who has seen some "stuff." Most of us who saw "stuff" don't talk about it openly, except to others who were there and then usually under duress … or drunkenness.

If you ever run across anyone who claims to be a veteran and starts spouting a lot of macabre sea stories right off the bat, question him closely. Chances are you will find that he spent most, if not all, of his tour as a "Remington Raider" (clerk typist) or as a lifeguard at some beach. My friend Dennis Kawalek talked to a homeless veteran one day on the streets of San Francisco. Dennis began to try and pin down this fellow about his experiences in Vietnam. One thing, the guy seemed a bit young to have been there. Another was that he claimed to have been an U.S. Air Force F-16 pilot. That was impossible because the F-16 was not yet even on the drawing boards during the Vietnam

War.

About 2004, I personally talked to a morbidly obese young man in San Francisco. He claimed to have been a Navy Seal and to have been a POW in Hanoi. He claimed that he had "escaped and evaded" all the way to Saigon to rejoin his unit. Funny thing about that is he would have had to travel almost the entire length of both North and South Vietnam to do so and would have had to pass at least 100 U.S. military bases or outposts along the way. Also, I know that anyone who had ever been a Seal would have never let himself get so grossly obese as this man was at his young age.

In their bar, the two sergeants at Pitsanoulk introduced me to Courvoisier, a fine French brandy. I drank at least one full bottle of it while I was there. I liked it so much I persuaded them to sell me a bottle from their stash. It has been my favorite brandy ever since.

The second day, after a routine morning of training Thai troops, we went to the sergeant's mess for lunch. I liked this place because even way out here in the boonies they had fresh dairy – milk, butter and ice cream. Even at Tango we didn't get fresh dairy. For dessert the sergeant grabbed a round two-gallon cardboard container of strawberry ice cream from the freezer and started flipping out portions with his combat knife.

I couldn't help but notice the knife. It was one I had seen frequently in magazine advertisements as the "Black Forest Knife." The accompanying picture usually showed the knife piercing a coin, and the copy stated that this knife would penetrate a coin without breaking off the tip. What caught my eye about the sergeant's knife was that the point *was* broken off. So I queried him, "Hey, Sarge, that is one of those Black Forest Knives. I thought you could stab it through a coin without breaking off the tip. What happened to yours?"

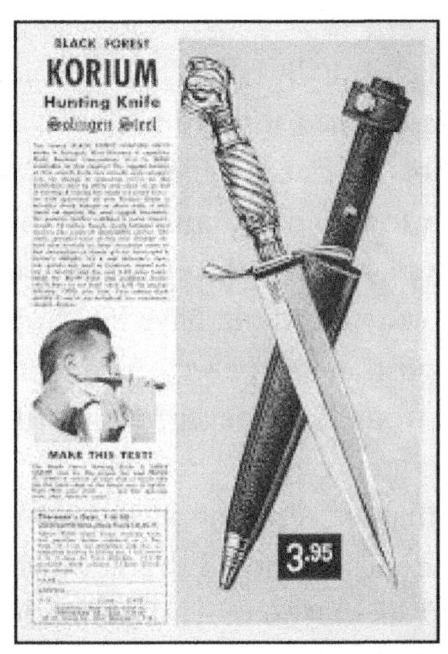

An old advertisement for the Black Forest Knife (WIKIPEDIA).

As he continued to dig out strawberry ice cream with his exotic knife, he told me an amazing story. He had been on a Green Beret team that was dropped far behind enemy lines in North Vietnam. The team's mission was to penetrate the defenses of a North Vietnamese Army regiment and to capture alive their Russian advisor. If they couldn't capture the Russian alive, their orders were to kill him. They did well. They penetrated the enemy's perimeter, and actually kidnapped the Russian without being detected. Unfortunately, as they made their escape, the alarm sounded, and they had to make a hasty retreat. The Sergeant did not say so, but implied that the drugged Russian at this time became excess baggage and had to be "disposed of." They were now in a dive situation fleeing for their lives.

The sergeant was running through the woods to avoid a sure death, as the enemy was closing in on his team. An enemy soldier passed close, so the sergeant hunkered against a tree. The enemy soldier paused with his back to the sergeant. The sergeant knew it was only a matter of seconds before the enemy trooper detected him and sounded the alarm.

The American grabbed the NVA soldier from behind, covering his mouth and thrust the sharp, pointy knife blade towards the soldier's throat, just like we have all seen in the movies. Unfortunately, the trooper had probably never been to the cinema, and chose not to act like the average movie victim. Instead of quietly taking a deep thrust to the throat, he struggled to escape. In doing so, he ducked his head forward, taking the tip of the blade of the Black Forest Knife into the top of his skull. Following through on his action in good military manner, the Green Beret sergeant gave the blade a sharp twist, which broke off the tip in his victim's skull.

With this same blade, the sergeant finished scooping out a second large portion of strawberry ice cream into another bowl … and handed it to me.

The evening diversion at this base was to go into the local small village and socialize with the sweet local working girls. My crew chief and I sat around the outside of a bar drinking beer, talking with them. They had an interesting diversion. They wrapped the strings of hydrogen-filled toy balloons with toilet paper. Then they lit the lower end of the toilet paper on fire and released the balloons. After a balloon rose above 20 feet, the flame would burn through the balloon and ignite the hydrogen, creating a nice little explosion to entertain us.

That is all that went boom that night.

MEANTIME, BACK IN THE REAL WORLD:

21 NOVEMBER, 1970

THE U.S. MILITARY WAS THWARTED IN THEIR ATTEMPT TO RESCUE AMERICAN POWS FROM A PRISON CAMP IN SON TAY, NORTH VIETNAM. EITHER THE ENEMY KNEW THE RESCUE ATTEMPT WAS COMING, OR THEY JUST HAPPENED TO MOVE ALL THE PRISONERS AWAY JUST BEFORE THE RAID. NO AMERICAN LIVES WERE LOST.

10 and 11 December 1970
H-64
Copilot for Captain Jerry Toman.
We flew one special mission each day near Louang Prabang area. (E-9).

22 DECEMBER 1970,

THE COOPER-CHURCH AMENDMENT TO THE U.S. DEFENSE APPROPRIATION BILL FORBIDS THE USE OF U.S. GROUND FORCES IN LAOS OR CAMBODIA.

28 December 1970

A crewman, Boumy Vongacgak, was killed by gunfire over LS-20 (F-9) while in an Air America Caribou.[3]

---

3 These incidents of crew deaths are taken from the chronology in the back of Jim Parker's book, *Timeline, Battle for Skyline Ridge*.

16

Bangkok

About my third STO, Gary, senior to me, worked with scheduling so that his STO matched mine. We were off together to the Siam Intercontinental Hotel to pursue Pan Am stewies. Pursue is perhaps too strong a word. They were not hard to catch.

Post card advertising Siam Intercontinental hotel. (AUTHOR'S SCRAP BOOK.)

On our first day at the Siam Intercontinental, we met a couple of Pan Am (PAA) lovelies at the swimming pool and began to learn the Pan Am system as it pertained to our needs. We learned that Pan Am Flight One landed about 10:30 p.m. We could expect the crews to arrive at the hotel bar after 11:30 p.m.

(WIKIPEDIA).

Pan Am was doing regular R&R flights from Vietnam. A crew would arrive in Bangkok one day and spend that night at the hotel. The next day they would do what they called an R&R turn-around. After delivering a load of GI's from Vietnam to BKK, they then got a three-day rest break. There might be four crews at the hotel on any given day. This was wonderful. We could meet them one day, get acquainted and then make dates for the three-day rest period. We took full advantage of this system. Get to know them, we did.

Giving them a little time for dropping off their luggage and freshening up a bit, we knew we could expect a crew of lovelies to arrive at the bar before midnight. We arrived early and hung out at the piano bar, lubricating our personalities.

One piano player was an attractive blonde Australian named Claire, whom we loved the best. The first thing she did whenever anybody sat at her piano was to ask, "What is your favorite song?" Then, whatever it was, she would play it beautifully. The next time that customer walked into the bar, she would break off and begin to play that person's favorite song. She got humongous tips for this.

Like the red-bellied ant vampire at Savannakhet, we waited beside the path the stewardesses would have to walk from the door to the big booth in the back of the bar to meet their friends. Then we would pick them off, one by one, grab them from behind, bite them on the neck, suck out their bodily fluids, and cast their spent little bodies onto the bar floor and await the next unsuspecting victim. Oh, no ... wait. It wasn't like that at all.

We would wait at the bar until the whole crew was gathered in the big corner booth at the back of the bar. We would then have the bartender serve a round of drinks to the entire crew—on us. One time we sent 17 drinks. This, of course, always got us an invitation to join the group. We would then make an instant snap judgment and do our best to sit beside an attractive stewie. (No problem there, they were all attractive.)

There was a nice nightclub attached to the hotel. After a drink or three, we simply asked one of the girls if she would like to see the late show and do a little dancing afterwards. We rarely got a refusal, and if we did, it was because the "ant" of our choice was tired, so we would make a date for the next night. We had a ball doing this. We met a wonderful array of beautiful young ladies.

Everything you have ever heard about stewardesses is true: they were beautiful, charming, educated, erudite, friendly and multilingual. And they liked to play. When we worked, we worked hard; when we played, we played hard. I know many of those girls felt the same way. Many times after one of these weeks on STO, I felt I must go back to work so I could rest up.

The first Pan American stewie I met after Susan, I nicknamed "Sweet Syn." After a night out on the town, we got together again the next morning. She told me she wanted to go across the street from the hotel to buy some jewelry. She said the jewelry shop gave a "Pan Am discount." Skeptical about the Pan Am discount, I agreed to accompany her. I thought I might save her some money. The store was about 50 feet square and must have had 20 display cabinets full of jewelry. Syn knew exactly what she wanted – a ring with three emerald cut sapphires set side-by-side. We looked for almost an hour. She found the three stones she wanted, but she did not like the setting. She found the setting she liked in another display case. She asked if they could transfer the stones to the setting she liked. Of course they could. Then we began to bargain on price.

I do not remember the exact numbers, but these are close approximations. When she asked the price for the finished ring, the shop proprietor said $349. She pulled out her checkbook to write a check for the full amount. I told her, "NO! Offer them $120." Shocked, she said, "I can't do that." I said, "Do it anyway." She did. They scoffed, and came back with $300. I told her to offer $160. She did. After several more offers/counter offers, I said to her, "OK, let's go!" I grabbed her by the hand and pulled her towards the door. She resisted. She wanted that ring.

As we neared the door, the shop proprietor said. "OK, OK, we'll take $189." I do remember that, whatever the final price was, she got the ring for about half of whatever they were originally asking. I accompanied the jeweler upstairs to watch him switch the sapphires from one ring and install them into the other ring so that he could not switch the stones for less valuable ones. When Syn took that ring back to San Francisco, it appraised for $999.

During our time hanging around the piano bar, a slightly older blonde made an effort to become friendly with us. After gaining our confidence, she trusted us enough to ask if we knew someone who could dispose of her husband for her. I guess our talk about flying "up country" and our reputation as CIA mercenaries had impressed her. She thought we might be available to do her dirty work. We assured her that we were simple pilots and not assassins for hire.

It was also while sitting at the piano bar that Bing Bengtson came up with a great bit of wisdom. He asked Gary and me what we were up to in Bangkok. Then he answered his own question with, "Oh, I know what you guys are doing. You are here to fuck yourselves to death – or die trying."

Bing was famous for his shaky hands. His hands would shake every time he did anything, but as soon as he grasped something, including the controls of an aircraft, his tremors ceased. He had no problem flying a helicopter. I learned later that this is called "intention tremors," a benign condition. We made it a point to tell new-hires to watch Bing closely — "If his hands start to shake—watch out! You will be in trouble." It gave them something to worry about.

For the next three or four STOs, we stayed in Bangkok. We got to know several stewardesses from several airlines and created for ourselves a nice social life. Some of the girls had enough seniority that they could bid their schedules to match ours. Gary and I linked up with Syn and her friend Lyn. The four of us got together several times.

Once we four had the intention for going to Ceylon (Sri Lanka) to see what it was all about. Just before we were to depart BKK, a civil war broke out there, so we changed our plans and visited the resort town of Pattaya Beach, stayed at a nice hotel and partied away a lovely week. I once counted up and realized that I managed to rendezvous with Syn at least one day for each of 11 out of 13 STOs.

In Bangkok (1971) We also met chicks from TIA charter airline.
(PHOTO BY GARY CONNOLLY)
The young lady on the right was Dory, a TIA stewie.
She is the one who gave me the nickname Bangkok Bill.

17

Cracked Impeller Housing

8 January 1971
H-86
FM Delacruz, R.C.
Long Tieng

I was assigned one of the newer H-34s, H-86, to fly upcountry and work around Long Tieng (*F-9 on map*). The machine was brand new and impeccably clean. It was perfect. I think this was its first trip upcountry. I anticipated an easy six-day tour.

For my first load out of the valley, I was given supplies to carry to a nearby hilltop LZ, just east and slightly south of Long Tieng. It was a safe area, occupied by friendly Thai troops. I checked my load chart for the max permissible load for the destination, loaded the helicopter and made a smooth rolling take off from the airstrip.

I found my destination – no great feat as it was visible when I took off, less than two miles away. I set myself up for a landing into the wind and approached the LZ. On short final, as I brought in full power to cushion the landing, the helicopter fell through, slamming into the ground just short of the center of the helipad. Fortunately, the helicopter's massive oleo struts and large balloon tires absorbed the hard landing.

Of course, I blamed myself and thought that I had just screwed up the landing and brought power in too late or whatever technique detail that I must have ignored. It was my first day back on the job after an STO. *Just rusty,* I thought. Delacruz unloaded the cargo, a load of rice in 100-kilo bags. I did a hover check, and all seemed OK. I shook my head thinking, *Boy, I really screwed that up.* I departed the LZ and returned to Long Tieng, still puzzled over why I had made such an awful landing, and vowing to be more careful on my subsequent landings. It would be so very embarrassing to crash on a routine re-supply mission, not to mention hazardous to health and life. I didn't want to go home in disgrace ... or in a body bag.

I loaded up with a second load of cargo. This time before I took off I did a hovering power check on the ramp, using max power available from the engine to see if there was enough power to hover before trying to take off. As I did, a really strange thing happened. When I brought the helicopter up into a hover, it reacted normally. I saw and felt the usual indications of a good engine, until I reached near maximum power, about 2800 RPM and 43 inches of manifold pressure. Then the engine simply seemed to run out of power. The aircraft settled to the ground, gently. Again I thought it was my technique, and so I executed the hover check over, admonishing myself to not be so sloppy with the controls, but ... HELL! ... the same result - a serious loss of power as I approached full power settings.

What the crap, over?

I told Delacruz that there was something wrong with H-86. We taxied over to the ramp and unloaded the cargo. Delacruz looked the helicopter over very carefully, checked the spark plugs, and the DMD (dual magneto distributor) the wiring harness, and anything else he could think of. He could find nothing. By then I doubted my own technique. Could it be that I simply screwed up the hover check two times in a row? Well, we would go out and do another hover check.

A pilot inspecting an H-34 engine with clamshell doors open.
PHOTO FROM ARCHIVES OF GARY CONNOLLY.

I cranked up, rolled the *now empty* machine out to the departure area, and did another hover check. This time it hovered just fine, no power loss at all, all the way up to hover power. That was the problem, as I was to soon discover, as the helicopter did not require as much power to hover when it was empty as it did when it was heavily laden. Satisfied that we had ample power, I returned to the loading area, loaded up yet another load of goodies for the troops on the hilltop, and made another rolling departure.

This time we almost died.

I set the helicopter up for a normal approach. When I brought the power up near maximum on landing, the helicopter engine again lost power. I began to fall out of the sky about 200 feet short of the LZ onto a hillside full of troops. The ground beneath me at this point was not flat enough to land on. This would have ended up as a crash for certain. There really was NO escape route.

I somehow did the right thing. In a situation like this, it is natural to try and increase power. Somehow I reacted against instinct, and reduced throttle. That gave me back some of the power I had lost just seconds before. I ballooned up, barely making the edge of the pad, landing somewhat short once again … and rather hard again … but intact.

I told Delacruz to throw out the cargo. We were going back to

Long Tieng to see what the hell was the matter with this helicopter. I was certain now that it was not simply a lack of technique or inattention on my part that was causing the problems. Some passengers wanted to fly down to Long Tieng but I told Delacruz, "NO!" I didn't want to take any chances with anyone's life but the crew's, now that I was certain that there was indeed something amiss with this machine.

We returned to the base. Delacruz once again looked over the helicopter carefully. Still he could find no problem with it mechanically. I told him to look once again. In the meantime, I telephoned Tango to see if anyone there could help me, but between awful telephone connections and possibly indifference, there was no help from that quarter.

Delacruz again found nothing wrong. Once again, I loaded the helicopter up, and took over to the take off area for hover check. This time, however, I had a heavy load on board, and when I pulled the machine up to a hover, it reacted perfectly, just as before, until I reached max power. Then I got the same reaction that I had already felt on the hilltop three times: a diminishing of the great roar of the Wright Cyclone engine, a great lessening of power and we again settled to the ground. I repeated this test several more times and got the same results. I did mag checks, power checks, re-checked to make sure my fuel mixture and carb air were correct; still NOTHING! I did a high-power burn-out to burn off any possible carbon build up on the spark plugs, even thought this was a nearly new engine and could not yet have had carbon build-up. No joy!

I took the helicopter back to the ramp and secured it. I told Delacruz to look it over once again, while I talked to some of the other pilots and tried to call Tango again. Finally, after an hour, Delacruz came up to me and said he had found the problem. The impeller housing of the supercharger had a hairline crack in it. This almost invisible crack in the metal that was nearly impossible to see with the naked eye. The crack would stay closed until near maximum power, at which time it would open, spilling supercharged air out to the atmosphere instead of channeling it into the carburetor to power the engine. At the times of highest demand, my horses were escaping the corral. I felt that the helicopter was safe to fly home, so we saddled up and returned to Udorn without event.

The Wright 1820-84A 9-cylinder radial was a good engine, but it was used up at only 600 hours. The first 50 and the last 50 of those 600 were high-caution periods. Those were the time frames in which we could expect most of our engine failures.

I had a couple of other routine engine failures during my time at Air America, but I do not recall ever having an outright engine failure. They always gave me some warning in the form of chip lights or instrument indications before deteriorating slowly. I was always able to land before one abruptly quit.

My friend Dennis Kawalek must hold the world's record for engine failures. Between USMC Vietnam and Air America Laos, he claims to have had 23 engine failures. No wonder he quit flying to become a storekeeper after Air America

In most aircraft piston engines, oil is circulated by an oil pump. As the oil passes the lowest position in the engine, it passes over a magnet. If a piece of metal catches on that magnet, it completes a circuit that lights up a warning light in the cockpit. Many times water or carbon build-up will complete the circuit too, but when a light comes on it is always prudent to land and check the sump for metal bits. A few small shavings are allowed, but they mean "WATCH OUT!" Large bits mean an engine is finished. Chip warning lights were almost never caused by big pieces of metal with these engines; until they were.

I made it a point to ask my crew chiefs if they could swim. If they said yes, I briefed them that in case of engine failure and the availability of water, I would prefer to ditch the helicopter rather than try to land in trees. I grew up playing in water. I was an expert swimmer and an experienced SCUBA diver. I had earned a merit badge in Navy flight school for my prowess in – and under – water. I would much rather take my chances at drowning versus being crispy-fried in a burning H-34. Most of the time there was no water nearby, so the idea of ditching was moot, but when there was a river or a pond nearby, I registered it in my memory. Just north of LS-272 there was a beautiful waterfall that spilled into a huge pond. I'll bet there is a big resort condominium complex there by now.

28 January 1971
H-80
My copilot is Fiorillo.
One special mission near PS-22 (O-19).

29 January 1971
H-80
Copilot for Captain Leon LaSchomb.
One special mission.

MEANTIME, BACK IN THE REAL WORLD:

30 JANUARY 1971 TO 6 APRIL 1971

OPERATION LAM SON 719 IN VIETNAM BEGINS. IT IS AN ALL SOUTH VIETNAMESE GROUND OFFENSIVE; 17,000 SOUTH VIETNAMESE SOLDIERS ATTACK 22,000 NVA INSIDE LAOS IN AN ATTEMPT TO SEVER THE HO CHI MINH TRAIL. DUE TO A LACK OF SOUTH VIETNAMESE AGGRESSIVENESS, THEY FAIL TO PUSH AHEAD AND GIVE THE NVA TIME TO REINFORCE AND REGROUP. THE SOUTH VIETNAMESE ARMY IS PUSHED BACK AND IT SEEMS THAT VIETNAMIZATION MAY BE FAILING.

31 JANUARY 1971

APOLLO 14 IS LAUNCHED AND ORBITS THE MOON 34 TIMES. ASTRONAUTS SHEPARD AND MITCHELL MAKE MANKIND'S THIRD MOON LANDING BEFORE RETURNING

TO EARTH. FIRST 6 IRON GOLF SHOT ON THE MOON.

7 February 1971
H-74
FM Ajero
My copilot is Jim Sweeney.
Two special missions near Savannakhet (K-16).

10 FEBRUARY 1971

LARRY BURROWS, AN ENGLISH PHOTOJOURNALIST, WAS BEST KNOWN FOR HIS PICTURES OF THE AMERICAN INVOLVEMENT IN THE VIETNAM WAR, DIED WITH FELLOW PHOTOJOURNALISTS HENRI HUET, KENT POTTER AND KEISABURO SHIMAMOTO, WHEN THEIR HELICOPTER WAS SHOT DOWN OVER LAOS. AT THE TIME OF THE HELICOPTER CRASH, THE PHOTOGRAPHERS WERE COVERING OPERATION LAM SON 719, A MASSIVE ARMORED INVASION OF LAOS BY SOUTH VIETNAMESE FORCES AGAINST THE VIETNAM PEOPLE'S ARMY AND THE PATHET LAO.

-WIKIPEDIA

12 February 1971
H-87
FM Lannan
Copilot for Dennis Kawalek. Four special missions near L-11 (N-18 on map).

February started the beginning of "Smokey Season." The indigenous people of Southeast Asia area practiced "slash and burn agriculture." There were times when the visibility approached that of a mild fog. You had to know your terrain to navigate around in this dense smoke. When we asked the Air Force bomber pilots who flew over to bomb North Vietnam how far up they had to climb before they got out of the smoke, they responded, "We don't."

18

We Move Out of Chai Porn Hotel to Better Quarters

After about six months in the Chai Porn Hotel, Gary and I agreed to move out of the hotel and to rent a house together. Dennis Kawalek lived in a nice little bungalow in a secure compound, and he knew of a two-bedroom house coming available. We rented it. The big factors were that it was cheaper than the hotel, and it was much nearer the base. We each had our own room, and we had a kitchen. Gary's favorite snack was to skewer a frozen hot dog on a fork, run it around inside a bottle of mustard, and eat it like a Popsicle.

During the time of our living in the hotel, there had been quite a flood, and for a few days it was impossible to get to the base by vehicle. The company actually sent up a helicopter commute service from the city fairgrounds to the air base for a few days to get employees to work, so we thought it better to be closer. Plus, we shopped at the base shops and went to the company club for most of our recreation.

Another advantage of living in a compound versus a single house was that it was secure. There was a pretty severe theft problem for those Americans living on the economy. It was not unheard of to have someone come home from a trip and find everything gone from his home, right down to the wall sockets and fans from the ceiling. A Thai police major owned the compound that we lived in. The word was out that he was a badass so we expected no problems.

AIR AMERICA LOG ★ エア·アメリカ·ロッグブック

Flooded AAM hangars.

UDORN

UDORN FLOOD

The city of Udorn, in central Thailand, recently suffered one of the worst floods in its history; the magnitude of the catastrophe can be seen by the pix on these two pages.

AAM's Base suffered virtually no damage as a result of the high water since only relatively small areas of the Base were flooded to any depth.

The salient point was the great damage the flooding did to roads leading to the Base from downtown Udorn and from the surrounding countryside which seriously impeded the ability of AAM employees to get to work. Result was a slight production slowdown for about three days.

To meet flight schedules and make sure certain key personnel could get to work, UH-34 helicopters were pressed into taxi service. Centrally-located areas in Udorn, which were known to be high and dry, were preselected as rendezvous points for those employees to be air-taxied to our Base.

The most frequent comment heard was: "This IS the way to come to work!"

* * * * * *

The memorandum reproduced below is self-explanatory — ED.

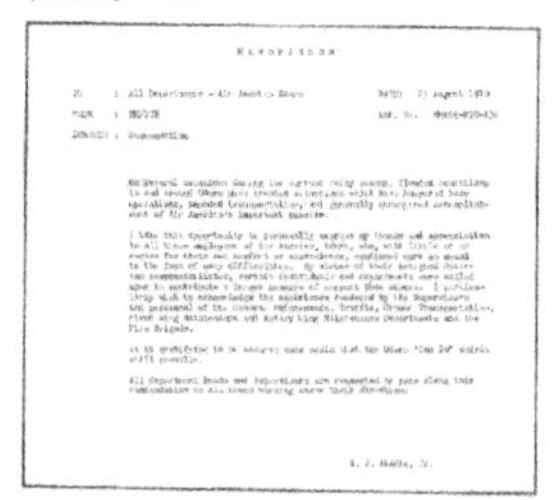

Biking and pumping in front of AAM's Traffic Terminal.

Downtown Udorn's Sowadee (Welcome) Circle — with double arches — completely surrounded by flood waters.

"YOU CANNOT FLY WITHOUT SUPPLY"

Pictures of a flooded Udorn from an AIR AMERICA LOG

One night I heard a commotion at the other end of the compound. I came out of my house in time to see two men running pell mell for the gate with several Thais in hot pursuit. One got away; the Thais caught the other. They pushed the man down and held him until the police major arrived. There was an awful lot of yelling and hysterical activity, but my understanding of the Thai language was minimal. Finally, the major showed up, and the whole cacophony of wailing and yelling and screaming began all over again.

Dennis Kawalek's Thai girlfriend (and future wife), Dang, arrived from her job as a waitress at the Air Force base officers' club. I asked her what was going on. She told me that they were trying to decide whether they should just shoot the *quemoy* (thief) right then and there, or take him to the police station and book him. Finally, they decided to take him to jail. There were too many witnesses.

Air America Pilot Beats Up Thief

A second-hand story is about the Air America pilot, Dave, who had already been robbed blind twice. He came home one day to find a third robbery in progress. Dave was so angry that he grabbed the *quemoy* and beat him severely. Then he locked the thief in the closet underneath his stairs and got in his car to go fetch the police. Before he could start his engine, he became so angry that he went back into his home, dragged the perp out from under the stairs, and beat him up again. Then he got into his car, once again going to the police. About half way to the police station, his anger got the best of him. Incensed, he turned the car around, went back home, dragged out the perp and beat him up a third time before finally handing him over to the police. Three robberies; three beatings. That seems to even out.

Fuck You Lizards

In our new compound lived large reptiles much like Gila monsters from the southwest U.S. These nocturnal creatures sometimes lived in the eaves of houses and preyed on small birds as they looked for roosting places about sunset. I am sure they ate rodents and geckos, too. They make a peculiar sound. From across the compound you might hear someone yell, "Fuck you!" At first it was startling to hear someone assault you verbally in this manner, especially in the darkness. You did not know who it was and why they were insulting you. Then we all learned it was the fuck-you lizards making their normal squawk.

19

STO Beirut

Gary and I took a tip from our friend Paul. On a prior STO he had gone to Bangkok and found the action slow. He boarded the midnight Pan American Airlines flight to Beirut, Lebanon, with a stopover in Tehran. We took the precaution of bringing pre-paid tickets with us. We could buy them cheaply and turn them in for credit, should we not use them. We went to BKK; things were slow. We decided to go to Lebanon for a few days.

We boarded PAA Flight One, westbound. There were not many passengers at that time of night and most of them were asleep. We had the entire Boeing 707 crew of stewardesses all to ourselves, cornered in the back galley of the plane. We dazzled them with our brilliance and baffled them with our bullshit. By the time we got to Beirut, we had dates.

We got our usual discount at the palatial Phoenicia Intercontinental Hotel where the Pan Am crews stayed. It resembled a lavish Roman-style villa with a huge pool surrounded by high, fluted columns. Down in the bar in the basement, we could have a drink and watch people swim in the pool through a huge window that created the pool wall at the deep end. This was great for girl watching. This was before Beirut had all the internal difficulties.

"This hotel became a battlefield in the Lebanese Civil War in 1975-6, during fighting known as the Battle of the Hotels, and was left a burned out ruin. It was abandoned for nearly 25 years until the late

1990s, when Mazen and Marwan Salha, Najib Salha's sons and members of the board of directors of SGHL, decided to restore the hotel."

-*Wikipedia.*

Gary and I took our dates to the big casino on the other side of the bay from the Phoenicia Hotel. The casino included a huge theater with a spectacular show. I remember horses prancing across the stage and dolphins jumping around in the water-filled orchestra pit. The casino itself was a bit too staid, not at all like those we were used to in Lake Tahoe or Stateline, Nevada, with loud music and free drinks. Neither of us was a gambler. We really didn't understand the games, so we did not play.

After the show, I was walking my date back to the hotel on a sidewalk next to a high brick wall. I decided I wanted to kiss her, so I led her through a cleft in the tall wall beside the sidewalk. We discovered we had stumbled into a graveyard. Oh, well, time to perhaps enact yet another fantasy. That was not to be.

About 90 seconds after we entered the cemetery, we were accosted by several Lebanese Army troops with machine guns who demanded to know what we intended to do in their cemetery. There was a bit of a language barrier. I feared we might be arrested and thrown into a Lebanese jail. My date saved the evening when she threw up her hands and exclaimed, "Honeymoon! Honeymoon!" It seems the soldiers understood that, for they let us go back out to the street with what sounded like an admonition for trying to be sneaky.

The next day, our crew of lovelies departed to the west. Gary and I took a guided tour of the ancient city of Baalbek. We found these ancient ruins most fascinating.

After two more days in Beirut, we caught a Pan Am Flight Two back to Bangkok. From the Air America office at the Don Muang International Airport we called the Udorn operations. I was on the schedule for standby at 5:30 the next morning. I had to be there. It was not in our mentality to let pleasure interfere with our work. I never once called in sick. We had a problem. We got back too late to catch the Thai Airlines flight back to Udorn. Oh, well, we'll catch the night train.

We looked out the window of the airport just in time to see the night train passing by. CRAP! Now what are we going to do? We had to get creative to get back to Udorn on time.

We hired a driver with a car to drive us home. Gary did not have to work the next day, so he volunteered to ride up front with the driver and watch him and help navigate. Gary insisted I lie down in the back and try to get some sleep. We had quite an adventure getting home. Gary said the driver got lost a couple of times, and Gary had to correct him. Trying to read the Thai traffic signs for the 250-mile journey was challenging, and the driver did something that many Asian drivers do, something that we never could figure out: He would turn off the lights of the car while driving down the highway at 50-plus miles per hour. Shit! There were water buffalo larger than our small Japanese car wandering about on the highway. We could have been killed in a collision with one of those, but Gary could not convince the driver to leave the lights on.

We made it back to our adjacent studio apartments about 15 minutes before my standby time started. I just had time to dump my suitcase, pack for another working trip, and get my uniform on when the mini-bus came to take me to work. As stand-by pilot, I was called to work because someone else had dropped out sick. I remember that, in this case, all I had to do was deadhead to Savannakhet. I was able to catch a couple hours sleep en route and then go to the company hostel and go to sleep for a few more hours. I would have been able to fly anyway, had it been necessary. We were young and full of energy.

Each Thursday the Rendezvous Club held a Mongolian BBQ poolside at the company compound. I loved that spicy food. Whenever I was at home, I made it a point to attend. One night I had an extra-large serving and chased it down with a lot of Scotch whisky. The next day I was deadheading to Savannakhet, napping on the troop seats down below in the H-34. Putrid gas badly needed venting from my body, so I let rip a huge fart. Both pilots flying the helicopter exclaimed how dreadfully awful was the stench wafting up through the cockpit.

I often went to work with a hangover. We disregarded the 12-

hour rule regularly, but only in the privacy of our homes and in the company of trusted friends. Rumors had it that there might be a snitch or two among our fellow pilots who might tell management if we were drinking after hours. There could have been some junior pilots who might snitch on us to gain seniority. The only fellows I really trusted were the ones I had flown in Vietnam with – Gary, Baiz, Morris, Howell, Koeppe, Ruck and a few others.

Sometimes after I came home alone, I would continue to sit and sip Scotch in private, too, but I never went to work with even a hint of being drunk.

20

Multiple Malfunctions

"You can always tell a helicopter pilot in anything moving: a train, an airplane, a car or a boat. They never smile, they are always listening to the machine and they always hear something they think is not right. Helicopter pilots fly in a mode of intensity, actually more like 'spring loaded,' while waiting for pieces of their ship to fall off."

-Author unknown.

Flight Details not recorded.

One day I had multiple problems with an H-34. It proved out to be quite challenging and amusing at the same time. I was working Long Tieng, hauling stuff and folks around the local hills, mostly south of the Long Tieng bowl. I began to have a little problem with the throttle of the helicopter. The motorcycle-type throttle on the collective began to try to increase the RPM of the engine all by itself. This is called "roll-on throttle." It really wasn't a big problem; it just meant I would have to monitor the RPM more closely, keep my left hand closer to the throttle and be ready to roll it back once in a while. The problem began to get worse and worse throughout the day, and it got to the point that I could not let go of the throttle grip at all. Soon I had to actually hold a constant backwards pressure on the advancing throttle to keep the RPM from increasing.

Then I began to get a little upload on the collective, the lever that increases the pitch of the blades and causes the helicopter to climb (if the throttle is correspondingly increased to overcome the increased drag). Now I had to keep a little down-pressure on the collective to keep it from rising. That was not a problem either, as I had the roll-on throttle to deal with anyway. These small problems were – so far – just a bit of a distraction.

Pressing on, I landed to load the helicopter. I turned on the stick trim so the control stick would stay put when I let go of the stick. I reached across with my right hand to operate the radio, since my left hand was tied up with the collective and throttle. When I did, the stick started to fall over, quite contrary to what I expected. The stick trim seemed to work most of the time, but occasionally seemed to have a mind of its own, and would intermittently decide to stop working. When that occurred, the stick would start to fall to one side or the other. The helicopter would have reacted accordingly. Now I couldn't let go of the stick either, except for short periods, and then I had to be in stable flight – and be quick about it – so the helicopter would not do any fancy aerobatics. I decided to press on. I was having a great day, and nothing was happening so far that was out of my control. I continued flying supplies to the troops on the local hilltops.

As if that weren't enough, I began to lose the hydraulic boost for my tail rotor. Sometimes it would work, and sometimes it would not. When it decided to stop working, I would have to push hard on the rudder controls. Then the boost would kick in, causing me to over-control and wiggle my tail on final. This caused me to lose the normal smoothness that I usually demonstrated. After this last system malfunction, I finally decided that I'd had enough, and the helicopter or God or somebody was trying to tell me something. I realized that I would be in a bit of a tight spot if I should have any further system malfunctions, such as an engine failure. In this situation, I could not ever let go of any of the controls at all. This made it very difficult to activate any switches, change any controls, or to change radio frequencies to make my quarter-hourly position report.

I decided it was time to return to Udorn. I called Oscar Mike and told him I was headed home with multiple system malfunctions and gave an ETA. The flight home was uneventful. I filled up several lines on the maintenance sheet describing the various things wrong with that helicopter.

The next day I was off duty. I went to the company compound for a late breakfast and to check my mail. Beside the pool was an American-type snack bar where one could order the usual things like burgers, fries, shakes, etc. I had been out late the night before, and it was just after 11:00 a.m. when I sat down at the snack bar. When the Thai waiter came to take my order, I ordered a breakfast of ham and eggs with toast. The waiter was gone for about five minutes when he came to my table and most apologetically said that he was sorry, but it was too late to order breakfast. I asked him to check with the cook, and to see if he couldn't make me ham and eggs anyway, as I was the only customer at the time. He was gone for about five minutes again, and even more apologetic when he returned to tell me "No, the cook cannot make a ham and eggs breakfast."

I asked the waiter for a lunch menu. Right there on the menu under sandwiches was "ham and egg sandwich." I reasoned that if he could make a ham and egg sandwich after 11 a.m. the cook ought to be able to make me a ham and eggs breakfast. Once again I sent the waiter to the cook to explain my reasoning and to request a ham and eggs breakfast. No luck; the cook would not budge.

I resigned myself to having lunch instead of breakfast. But then I said to myself, at least I can have a ham and egg sandwich. Then I had a clever thought. I called the waiter over, and I ordered a "ham and egg sandwich without any bread, toast on the side." In a few minutes, my "ham and eggs sandwich, without any bread, toast on the side" breakfast arrived. It cost me five cents more than what I originally wanted.

Merry-Go-Round

FEB. 18, 1971 / S.F. Chronicle

With the CIA Agents In Southeast Asia

Jack Anderson

THE POPULAR impression of CIA men in Southeast Asia is of lean-faced James Bonds talking in whispers to Indochinese beauties in dingy bars or of bearded guerrilla experts directing Meo tribesmen in the Laotian jungles.

The real McCoy, more often, is a rumpled civil servant going to lard, who worries about when his refrigerator will arrive from the States and plays bingo on Tuesday nights.

This is the unromantic picture that emerges from an instruction sheet handed to CIA pilots leaving for Udorn, Thailand. The CIA uses a front called Air America to fly missions out of Udorn over Indochina.

★ ★ ★

INSTEAD OF pressing cyanide suicide capsules upon new recruits, the stateside briefer slips them a bus schedule for CIA personnel between Udorn's CIA compound, schools und banks.

"A bowling alley in Udorn has league bowling," the CIA confides to its pilot-agents. Their wives are given such hush-hush CIA tips as "water should be boiled three to five minutes prior to drinking, but it is safe for cooking and washing dishes if it is brought to the boiling point."

Wives are also advised to bring "plenty of sheets and pillow cases" and "chinaware, tableware and kitchen utensils."

Other confidential information provided CIA agents includes the intelligence that "Thai mattresses are normally extremely hard and bumpy" and that "shopping is generally done by the servants due to the early hours (6 a.m.) one must shop to insure getting fresh products."

The cloak-and-dagger boys are told they will have a supermarket, swimming pool, free movies, the "Club Rendezvous" (which doubles as a chapel on Sundays) and bingo on Tuesday and Saturday nights. The CIA bars are called The Pub and the Wagon Wheel and shut down at midnight. The same humdrum life style can be found at such CIA outposts as Vientiane, Laos, where CIA men usually live with their families in villas and dine at the town's few French restaurants.

★ ★ ★

BUT IF the CIA living conditions are vintage suburbia, some of the missions are dangerous. The CIA pilots fly supplies to CIA-backed Meo tribesmen in Laos hinterlands. There are also more hazardous missions, such as flights along the Red Chinese border and ammo deliveries to tiny airstrips in communist-infested country.

Footnote: Much of the recruiting for CIA pilots is done out of a gold-carpeted office in downtown Washington with "Air America" on the glass doors.

Basic pay is $22.98 an hour for captains, $13.93 for first officers, with bonuses for special "projects." A top CIA pilot can make as much as $100,000 a year flying high hazard missions. In addition, station allowances run $320 a month at Saigon, $215 at Udorn and $230 in Vientiane.

Columnist Jack Anderson article in San Francisco Chronicle, 18 Feb. 1971.
FROM AUTHOR'S SCRAPBOOK.

24 Feb 1971

H-79

FM Stickler, Steve

Copilot for Captain Joe Lopes.

One special mission near Louang Prahbang.

27 and 28 February 1971

H-57

FM Livereza

One special mission each day out of L-11 (N-18 on map).

21 March 1971

Captain Benjamin A. Franklin died as his Porter stalled and crashed into a mountain near LS-272 (G-10 on map).

2-6 April 1971

H-74

FM Wade

Copilot for Captain Sandy Sandt.

Eight special missions this week near PS-44 (O-19 on map).

Top of the music charts from April 17 to May 22, 1971,
and what became the number one song
in the USA for the year was
"JOY TO THE WORLD" by Three Dog Night

17 April 1971

H-70

FM not recorded

Udorn local flight.

Captain's proficiency check with senior instructor James McEntee.

21 April 1971
H-80
FM F. Delacruz

I fly nine specials near LPB, with Jess Hagerman as my copilot.

29 April 1971
Captain Harry Mulholland dies in a mid-air collision with U-17 near LS-63

24-26 May 1971
H-47
FM Latloi
Copilot for Captain Tony Byrne.

We fly eight special missions these three days near Savannakhet (*K-16 on map*).

31 May to 2 June 1971
H-87
FM Legaspi

We fly four specials during these three days in Pakxe area. Copilot not recorded

2 June 1971
H-87
FM not recorded
Byron Ruck is my copilot.

I lead two specials. I made no note and have no recollection of details. These missions must have been routine missions. Near LS-11 and PS-44.

2 June 1971
Capt. Herbert Clark and his crewman Trilit Thuttanon die when their aircraft is hit by ground fire near LS-32 (Boung Lam). Three other crewmen bail out and survive.

5 June 1971
H-62
FM Leveriza and Mondelo. (Trainee)
Copilot John Ferris. We fly one special this day near Nam Lieu.

15 June 1971
H-79
FM Manalo

"LORAN C-123 bombing test" (?) This is another mystery entry in my logbook. I have no memory of what we were doing testing bombs with C-123s or why I was involved.

22 June 1971
H-57
FM Pichet

I fly copilot for Captain Frank Stergar. One Special Mission near Savannakhet.

24 June 1971
H-69
FM Manalo

My logbook shows we had a chip light at LS-279.

I celebrate the completion of one full year with Air America.

21

STO June 1971

I took my STO this month (8-14 June) and returned to my hometown of Sonoma, California, to attend my ten-year high school reunion. It was great fun to be with my classmates and brag about being an air charter captain with Air America. I doubt any of them knew what Air America was at the time. I won the prize for travelling the longest distance to attend–from Bangkok. I actually began this journey from LBP, far northwest Laos.

Because I travelled to California for my reunion, Gary went solo to the Siam Intercontinental Hotel. He tied up with Claire, the piano player, and spent a few delightful nights with her. She also read cards and offered to read Gary's future. It seemed like fun and games until suddenly she blanched and quickly swept the cards from the table.

Gary asked her, "What?" She said, "Oh, nothing." He reported seeing the ace of spades in the mix.

Gary died in the helicopter crash about four years later.

MEANTIME, BACK IN THE REAL WORLD:

15 JULY 1971

PRESIDENT NIXON ANNOUNCES THAT HE WILL VISIT COMMUNIST CHINA IN 1972.

20 July 1971

H-85

FM Freddy Alor

"Raven engine recovery." No details in my logbook or memory.

26 JULY 1971

APOLLO 15 ASTRONAUTS SCOTT AND IRWIN MAKE THE FOURTH MOON LANDING.

2 AUGUST

U.S. ADMITS TO HAVING 30,000 CIA-SPONSORED IRREGULARS OPERATING IN LAOS.

AUGUST 1971
THE SONNY & CHER COMEDY HOUR MAKES ITS DEBUT ON CBS.

I mark my 28th birthday this month. A senior pilot describes Gary and me as "budding millionaires."

22 August 1971,

H-44

FM DeCosto

Copilot for Captain Emmet Sullivan. We fly three special missions this date near Pakxe.

23 August 1971

H-85,

FM Delacruz

I fly a special mission with my good buddy Gary near Pakxe.

2 September 1971
H-53
FM Delacruz

I carry some VIP this date. I believe this is the date I carried General Vang Pao from LS-20A to Vientiane.

22

Collective Jam-up

13 September 1971
H-87
FM Baccay

I was enduring a captain's route check ride with senior instructor McEntee, working out of Lima Site 272 (*G-10 on map*). We had just picked up a load of rice for the landing zone of a nearby village, located through a small cleft in the hills not a half mile away. This kind of work was as routine as it could get—easy, short legs in a peaceful, friendly area.

As I came in on short final approach to land at the LZ, the collective of my H-34 would not bottom out. I could not reduce power all the way as I needed to for landing. Something was jamming up the controls. This was uncomfortable. I didn't know if the condition could get more serious if I tried to make the short hop back to 272. I held full back on the cyclic stick and jumped on the toe brakes as hard as I could without causing helicopter to nose over. I rolled past the center of the zone, right up to the very edge of it, and shut down.

Capt. McEntee asked me why I didn't go back to 272. I explained my concern; "I don't know what's wrong with this helicopter and do not want to take a chance that it is a serious control malfunction."

Flight mechanic Baccay and I made a quick inspection of the helicopter. As I inspected the collective mechanism, I found that a small bolt had fallen from somewhere above and settled into the collective detent, which blocked the collective from going all the way down. This small object was what had kept me from being able to reduce power as needed for a proper landing. I picked the screw from the detent and solved the problem. We went back to work.

It was a simple thing, but it might very well have become complicated. I didn't know at the time what was wrong and I did not want to take the chance of flying back to LS-272 without knowing. I guess that McEntee agreed with me because he never said another thing after that. (Sometimes I wonder if McEntee had placed that screw in the detent just to see how I would handle it.)

As I returned to LS-272 from that village, I saw a huge black snake slither across the runway. It must have been 10 feet long and eight inches in diameter.

23

My Longest Day

Almost stranded overnight

13 September 1971
H-87
FM Baccay

Right after the route check with Capt. McEntee.

On my last load out to the east from LS-272, carrying cargo to LS 266 (*H-11 on map*), the weather turned to crud. As we came to the last pass to cross back into the LS-272 valley, we were blocked by low clouds, making it impossible to continue westbound. I circled around for a few turns, hoping that there would be a hole, but the weather just got worse and worse. At one point, I wiggled my tail a bit trying to get a better view of things. My being out of balanced flight for a few seconds allowed the heavy rain to enter my pilot's window. The effect was like having a bucket of water tossed into my face.

Not only was the weather getting worse, I was now getting low on fuel. This had been my last trip for the day, and I planned my fuel to get me back to 272 before quitting time. There I would refuel for the short return trip to VTE. I knew that some of my friends had been in similar situations before, and the only reasonable thing to do was land at the nearest village and request to spend the night there.

My imagination ran the gamut of thoughts while considering this move. The village should be friendly. I visualized myself spending the night at a welcoming celebration around a huge community fire, roasted pigs, native drinks, and the chief's daughter(s) at my disposal. This was a welcoming thought. On the other hand, my experience in Vietnam brought another scenario to mind. I could see myself bound up, naked, trussed up like a pig on a long stick, delivered to the local communist cadre to become a prisoner of war – or worse. Not a pretty thought. I felt I could trust these simple people, but did I really want to take that chance?

We landed. Baccay walked into the village and talked to the headman. Sure enough, we would be welcome for the night. I did not relish the idea of sleeping in the "Sikorsky Hotel" should fantasy number one not occur. I had already called O.M. on the high frequency radio and told them of our predicament, so at least O.M. knew where we were. About that time, I noticed that there seemed to be a slight crack under the clouds that obscured our escape route. I told Baccay, "Let's make one more quick attempt to get out of here!"

I cranked up the engine in record time and blasted off, climbing to the west. I was right. There was a "sucker hole" that looked like we might make it through to LS-272. As soon as we entered the little arch of clouds above the pass, the whole valley opened up. We were able to continue westbound. Now we encountered another problem. The re-fueling crew at LS-272 was long gone for the night and I didn't have enough fuel to make it all the way to Vientiane. When I called on the radio to report my predicament, Hal Miller, in another H-34, came up with a solution. He was flying somewhere between the dam site and VTE, he told me he would bring a barrel of fuel to the little strip (LS-157, *F-13 on map*) just south of the dam site. I rogered that and headed south. When we arrived at that strip, a barrel of fuel awaited us, thanks to Hal. Baccay broke out the barrel pump, squirted that fuel into our tank, and we went home. We arrived at Vientiane well after dark, contrary to our usual habit of landing before sundown. Until then this was the longest day's flying I ever logged, eleven hours and 38 minutes.

On 13 Sept 1972, exactly one year later, I would have a day of 11:40 hours while flying Hueys. I had yet another day of 11:40 hours on 4 December 1972, three days before I quit.

FM Baccay refueling my H-34 at a remote site LS-157 after we escaped the bad weather east of LS-272. We got home to Vientiane well after dark.
FROM AUTHOR'S PHOTO ALBUM.

In a situation like this, we had to be careful about which fuel we used. I have photos of one batch of fuel somewhere north of Savannakhet, where all the barrels of fuel have been marked with skull and crossbones. Out-dated, badly contaminated, or both. We dared not ever use any of that fuel.

It is hard to see, but the little sign sticking up between the 4th and 5th barrels from the right has skull and crossbones on it.
DO NOT USE THIS FUEL! FROM AUTHOR'S PHOTO ALBUM.

Whenever we Remained Over Night (RON) at Vientiane, we sometimes visited Madame Lulu's Rendezvous Club at the airport. It was also called Lulu's Turkey Farm.

Madame Lulu's Turkey Farm, in Vietnam
where every girl was a gobbler.

24

Bizarre Cargoes

A few times working out of LS-272, I carried ludicrous cargoes. I once spent two days carrying cases of canned lobster Newberg to the villages. It was a donation from some U.S. company, and this food was so rich that it made the local people sick.

Another time, I carried literally tons of dried milk powder, only to learn later that the people could not eat this either because of lactose intolerance. The locals found that these inedible-to-them foodstuffs made great pig food.

I also learned that the people here historically ate their pork raw. That is they did until USAID introduced American Duroc pigs to cross with theirs to make them healthier and meatier. That part worked, but it also brought them trichinosis, which they did not have before. I never carried baby food as mentioned in the **oui** article. I mentioned the buckets earlier.

Another time I was flying in the LS-272 area, carrying many sacks of locally grown corn to a nearby 'ville. On the radio, I heard B.J. Singleton say the load in his Helio Courier included several pigs. I called him and joked over the radio, suggesting we get together. His pigs, my corn; we could go into the pig farming business.

A Helio Courier—COURTESY OF CAPTAIN ED ADAMS

B.J. had a little trouble with one load of pigs. The way the local people transported their pigs was to simply stick one in a gunnysack and tie the bag closed. As B.J. was flying along, a fairly large pig got out of its sack and began running around inside his tiny airplane. Frightened by the noise and the strange environment, it was trying to escape. As the pig ran to the rear of his aircraft, it upset the center of gravity (balance) of the aircraft. Even with full forward stick, the plane still wanted to climb. It started to slow down, and would have eventually stalled.

But then the pig changed direction and ran up to the front of the plane, underneath the instrument panel. B.J. then had the opposite problem. The pig under the instrument panel threatened to damage the aircraft's vital wiring and instruments. Worse than that, the balance of the little airplane was now way forward, such that even with full back stick, the airplane wanted to pitch head-down towards the ground. It was up and down and up and down for a while as the pig ran back and forth. Finally, B.J. managed to grab the pig as it ran past him. He wrestled it into a hammerlock and held it until he landed the plane with one hand. After that, B.J. made sure all the pigs in bags were securely hog-tied in addition to being tied into the burlap bags.

After a cargo drop, it was back to LS-272, get another 10 or 12 or 15 bags of rice, depending on the fuel load of my aircraft, and then back out to the next village for another delivery. Check the signal, make a cautious approach, land up-hill if needed, sit on the ground while the crew chief and locals unloaded the bags of cargo, and then back to LS-272 once again. I did this routine sometimes for five or six ten-hour days in a row. LS-272 was a good place to work because there was always lots of interesting, challenging work to do, but it was far enough away from enemy activity so as to not be very dangerous.

After a trip there, I would be pretty tired, but very satisfied that I had put in a good week's work and had gotten half my flight hours for the month.

17 September 1971
H-54
FM Phalowan

Another proficiency check ride around Udorn airport with senior instructor McEntee. I shot several full-down autorotations.

25

Our In-House Intelligence

At operations in Udorn, we had a very good intelligence office. A fellow named Jack B. was always on duty to tell us the latest intel and to take our reports should we have anything to contribute. We always checked in with Jack before a trip, and always reported back to him upon our return. The local Hmong troops had teams in the field and knew we could rely on the accuracy of their information much more than any we received in Vietnam. In Vietnam, we never knew who was on which side. In Laos, the Hmong hated the communists and were fighting for the very survival of their people.

Around this time I was working out of Long Tieng (*F-10 on map*) single pilot. I was tasked with delivering supplies to an outpost about 30 miles away. I remembered my brief from Jack that there was a reported .50-caliber machine gun on a ridgeline in the area, on a direct line between Long Tieng and my destination. The big gun had been reported months before and was already marked on my air chart.

After three round trips to the outpost, I began to wonder. That report of a .50 caliber machine gun was months old. There was really nothing in that area for the NVA to care about or to protect. I decided I would cut five minutes off my leg to the delivery point and fly over the reported danger.

As I passed right over that ridgeline, the gunner opened up. POP POP POP. The slow report of .50-caliber bullets passing by got my full attention. A .50-caliber bullet is as big as your thumb and can rip an aircraft apart in seconds. Sometimes they are filled with explosive and can be little hand grenades. I took immediate evasive action to exit the airspace above the big gun. Fortunately, this shooter was not well dialed in on us. He probably had minimal experience and never got any practice.

When I returned to Tango, I reported to Jack that the .50 caliber machine gun was still active. He did not ask me how I knew.

26

About Drug Running

To my personal knowledge, no Air America personnel ever ran drugs. None of my close friends nor I had any experience with or knowledge of Air America pilots carrying drugs.

The local people we carried considered opium to be one of their main medicines. There is no doubt we carried troops and civilians who carried small amounts of opium for their personal use. I can honestly say I never saw even that; but we never searched these people as they boarded our aircraft.

There were several Thai pilots who flew for Air America. Once in a while we might hear one of these Thai pilots calling departure from one of our bases at sunset when all our machines were supposed to be on the ground, but I never knew what they were up to. Were they perhaps flying clandestinely to further a mission? Perhaps they were picking up loads of drugs and delivering them to a depot somewhere.

We never knew. We didn't ask.

About 1994, I did meet a U.S. Air Force pilot who swears he watched as Vietnamese airport workers at the Saigon airport unloaded bricks of opium or heroin from Air America aircraft and transferred them to major airliners, but I did not personally see anything to support this.

This article by Jack Anderson, cut from the *San Francisco Chronicle* in 1971, seems to be a bit contradictory.

Merry-Go-Round

The Men Behind Laos Drug Traffic

Jack Anderson

A ROYAL Laotian prince and the Laotian Army commander have now been identified as the principal traffickers in the heroin used by U.S. troops in South Vietnam.

Furthermore, a Congressional investigation has confirmed our earlier allegations that the Central Intelligence Agency is involved in the Laotian heroin operations.

The investigation was made by Representatives Bob Steele (Dem-Conn.) and Morgan Murphy (Dem-Ill.) both members of the House Foreign Affairs Committee.

Steele is preparing a report that will allege CIA "Air America" aircraft have been used to transport the drug from northern Laos into the capital city of Vientiane.

It says, however, there is no evidence that the CIA had any official policy of letting its planes be used to move the drugs. Furthermore, it adds that the agency has now cracked down on the practice.

★ ★ ★

ACCORDING to the draft report, prepared by Steele for the House Foreign Affairs chairman Tom Morgan (Dem-Pa.), the deadly drug is transported from opium fields in Laos to the battlefields of South Vietnam in the following manner:

First the raw opium is hauled from northern Laos through Burma and into the Laotian town of Ban Bouei Sai, with former Nationalist Chinese soldiers - turned-drug smugglers riding shotgun on the shipments.

At Ban Bouei Sai, Laotian Army commander General Ouan Rathikoun supervises the shipment of the opium into Vientiane, using American - supplied planes.

Once it reaches Vientiane, the morphine base is processed in Rathikoun's labs into "Number Four" heroin, a pure grade of the deadly drug almost unknown in Southeast Asia until traffickers began turning it out especially for American troops.

The heroin operation is protected and abetted by Prince Boun Oun, inspector general of the realm. The prince gets part of the take.

★ ★ ★

ONCE PROCESSED, the heroin is flown into South Vietnam aboard military and civilian aircraft.

Some of the packages of the white powder are air-dropped near U.S. troop emplacements. Others reach the troops after being landed at outlying air strips or flown directly into Saigon's Tan Son Nhut airport.

With Vietnamese custom officials looking the other way, the heroin passes into illicit channels. The Congressman identifies South Vietnamese Premier Tran Thien Kheim as the man behind the corruption of the customs agents, but they stop short of calling him an outright trafficker.

The angriest language in Steele's draft report is reserved for U.S. diplomats who have failed to use their leverage to get the drug traffic cut off at its source.

Columnist Jack Anderson article from a 1971 *San Francisco Chronicle.*

FROM AUTHOR'S SCRAP BOOK.

Because the company started getting a lot of press about drug running, the company started checkpoints and inspections of all aircraft and personnel leaving Laos. One of our newly hired helicopter pilots did not know this. A big pot smoker, he soon discovered that he could buy a large amount of marijuana buds for little money in Laos. According to him, he filled a footlocker (his words) with highly potent MJ, and he walked right through the Company checkpoint with that footlocker full of buds under his arm. It did not register as heroin or opium, so nobody cared. He and his girlfriend intended to turn that collection of buds into hashish, but they did not know how. They boiled a lot of it in water and tried to smoke the dried residue, but found the boiling had ruined the effect. Only years later did they learn that the way to make hash out of marijuana buds was to beat them against the wall and scrape off the residual pollen—which is what hash is.

Follow-up

Three drug runners

After Air America, three of our guys were caught running drugs. Pilot X spent a few years in a Florida prison. I last saw him in Dallas in 1987 at the first Air America reunion I attended. He died a few months later, way too young.

Another of our helicopter pilots, Y, got caught running drugs into Texas from Mexico. He spent nearly 20 years in a Texas prison.

Yet another one of our fellows, Z, got arrested for running drugs over the border into Arizona. He got off only because he had his visor down, hiding his face, and the eyewitness could not positively identify him. This pilot said his lawyer's fees cost him most of his ill-gotten gains, but at least he stayed out of prison. While he was in jail awaiting trial, his wife cleaned out whatever was left in their safety deposit box and deserted him.

There are common threads here. We all needed the occasional adrenalin rush to make us feel alive, and we all had gotten used to making big money.

Fortunately, I avoided that genre of employment.

27

The Movie "AIR AMERICA"

In contrast to a popular (and stupid!) movie starring Mel Gibson, "AIR AMERICA" was a most professionally run air charter service. Every pilot had to have a commercial license. We had to hold and maintain a Class II flight physical. Each pilot had to have a type rating in any aircraft he flew, if that aircraft required a type rating (The H-34 did). Many of our pilots had the top-most aviator's rating, the Air Transport Rating (ATR). We had instructor pilots. We had routine proficiency checks and thorough route checks at regular intervals and annual flight reviews. We abided by FAA rules as to crew flight times and duty hours.

Our mottos were:

"Anything, Anywhere, Anytime, Professionally."

and

"YOU CALL, WE HAUL"

From Wikipedia:

"A 1990 American action comedy film directed by Roger Spottiswoode, starring Mel Gibson and Robert Downey Jr. as *Air America* pilots, during the Vietnam War, flying missions in Laos. When the protagonists discover their aircraft are being used by other government agents to smuggle heroin, they must avoid being framed as the drug smugglers.

> "The plot of *Air America* is adapted from Christopher Robbins' 1979 non-fiction book, chronicling the U.S. Central Intelligence Agency financed airline during the Vietnam War to transport weapons and supplies in Cambodia, Laos and Vietnam in the 1960s, subsequent to the North Vietnamese invasion of Laos.
>
> "The publicity for the film, advertised as a light-hearted buddy movie, implied a tone that differs greatly with the actual film's tone, which includes such serious themes as an anti-war message, focus on the opium trade, and a negative portrayal of Royal Laotian General Vang Pao (played by actor Burt Kwouk as General Lu Soong)."

In my opinion, there was as much truth in that movie "AIR AMERICA" as there was truth in Charlie Chaplin movies about cops – almost nothing. The only small part of the movie that held any veracity at all was the scene early on when the new pilot is headed out to his first landing. The two pilots fly around a limestone karst and come upon the tiny strip. The new guy says, "You want me to land there?" That is what a majority of the small strips were like. Most were dirt, 50-feet wide and 600-feet long. In that short distance, the strip might have a 20-degree dogleg in it, a small swale or a small hump in the middle. Some strips had both a dog leg and a hump or swale. Many times a Pilatus Porter or Helio pilot could not see the end of the strip once he sat down.

Many of the strips were steep. Once an airplane hit the ground, the pilot had to add full power to taxi up to the turn-around at the top of the hill or be stuck where he landed. The up-hill end of each strip was nearly always a flat area with room to turn around, so that the smaller airplanes could actually swing around and point their noses downhill before stopping. The reason for this were that there really was only one way to depart such a strip, regardless of the winds, and that was downhill. Another reason for this was so the pilot, who rarely stopped his engine, could make a quick get-away should anything scary happen, such as enemy troop rushing him, incoming mortars or RPG fire. Instances of this happening were rare, but not unknown. It was always better to be prepared rather than be captured or killed. The top, flat end of the strip doubled as the local helipad.

The pilots often had to buzz the field prior to landing to chase off water buffalo or children playing on the strip. More than once, a local person, ignorant about airplanes, died when he inadvertently walked into the propeller of a Swiss-built Pilatus Porter short takeoff and landing (STOL) plane.

To see videos of Air America pilots landing on these dust strips see bibliography,

1. Air America, "Flying Men, Flying Machines" by John Willheim.
2. Air America, "The CIA's Secret Airline."
3. "The Rescue of Raven 1-1," by April Davila. 25 minute video on youtube.
4. "Laos, The Forgotten War" www.youtube.com/watch?v=XB9oXpN2Owg

28

LS-118A

As a new captain, I worked the most northwest area of Laos at LS-118 *(E-6 on map)* for Customer Tony Poe, delivering supplies to remote hilltop outposts. CIA officer Tony Poe was infamous for his actions in his area of operations.

A former World War II Intelligence Marine, it was said that he operated his very own small mercenary army of local troops, and that he gave out bonuses for enemy troops killed. His way of verifying a kill was that the local trooper had to bring in the left ear of his kill. It was also rumored that he had married princesses of several of the local tribes to consolidate his power and leadership of the area, and, perhaps falsely, that he was the inspiration for Colonel Kurtz in the movie, "Apocalypse Now."

Sometimes, as I landed on some of these isolated outposts, all I had to do was extend my final approach a wee bit, and I would have crossed the Mekong River into the air space of Red China. I gave it some thought, just so I could brag I flew into Red China, but chose not to tweak the devil's nose in this case. Some cheap thrills are not worth the risk of spending years in a POW camp; or worse.

Another time I was supplying our post in that same extreme northwest area of Laos where a tiny finger of Burma sticks well into Laos. I was diverting a bit out of my way just to avoid flying over this tiny bit of Burma (*E-6 on map*). One trip, I decided to fly about five minutes across that finger, which I did without problem. I dared not do it again.

A few months later, one of our H-34 pilots got lost in that area due to bad weather. The visibility was terrible, and he made the mistake of going "on top" of the clouds. Stuck up there with no visual contact with the ground, he then (of course!) got a low-fuel warning light. Desperately in need of a safe haven, he spotted an airport below through a small hole in the clouds and descended to land – in Burma. He created quite an international incident. It was a couple of weeks before the Burmese Army got around to releasing him, fearing he was a spy. I believe they kept his helicopter. This pilot's name was Ben, and forever after he has been known as "Burma Ben."

Earlier on in my training phase as a new first officer, I had worked this area with a senior captain. He mismanaged his fuel and got us far away from LS-118A with minimum fuel. Worried about running out of fuel, he did something that to me was completely contrary to what I would have intuitively done. He flew low, skimming treetops all the way back to LS-118A. If the engine had quit under those circumstances, we would have plowed instantly into the trees.

I would have felt much more comfortable flying quite a bit higher, giving myself altitude to shoot a longer, more controlled autorotation should the engine quit. This, too, would give me time to get out a Mayday call. Perhaps he thought it better to not use fuel to climb up to a safer altitude. He never said. I never asked. The low-fuel light had been on for 20 minutes when we landed. Twenty minutes worth was supposed to be the amount of fuel left in the tank when the warning light illuminated, but we never trusted those warning lights to be reliable. I always managed my fuel so that I was on the ground before the 20-minute light came on.

Later, sitting around talking with my friends about this captain and several others, we all agreed that some of these senior pilots were not all that great as pilots; in fact, some were downright scary. We began to compile a list of the dangerous guys whom we dreaded to fly with and we gave them the nickname "The Fantastic Five."

There were about 150 helicopter pilots flying out of Udorn at the time.

29

We Move Again

About a year after our move to the compound, Mama-san, the major's wife, started to build a row of seven efficiency apartments behind our house. Gary and I took a look at them and decided we each wanted to have our own place. Three of our friends moved into the studios, too. It was the Chai Porn Hotel situation all over again, with open door policy – if you were home, the bar was always open, unless you were flying soon.

In order for Mama-san to rent to us, she requested that we pay a full year's rent in advance. We choked at hearing that until she told us that the rent was only $72.50 a month, so a year's rent was less than $1,000. We bargained with her not on the amount, but on some of the design features of the little units. Gary, for some reason did not like the fact that Thais liked to insert a transom window over each internal door of the house, so he bargained that away. There were a few other changes that we had made, which I don't recall.

As the Thai carpenters built these efficiency apartments, Gary and I would go over and watch them working. What amazed us was the way the Thai carpenter drove a nail into the dense teak wood. He would hold the nail between his first two toes and then hit it once hard, driving the nail all the way down on the first blow – no preliminary tap to get the range – pulling his toes out of the way as the hammerhead hit the nail. It was amazing to watch. The carpenters did not seem to have disfigured toes at all from any kind of learning curve.

One thing we didn't catch until it was too late. The walls between the units were paper-thin. Gary and I could carry on a normal conversation between our apartments without raising our voices. Fortunately, we were not often home at the same time. If we were, we were out playing together, so that neither of our privacies suffered from this. Once or twice he had a woman over, and I had to listen to various moanings and groanings. I'm sure he heard some of mine, too. We agreed that if one of us heard such noises from the other guy's side, the one hearing the noises would be quiet, so as to not scare the other's guest away.

One of the bennies not included with the apartment was a maid, however, one was available for a small monthly fee. I loved this. I could come home from a six-day trip, drop my suitcases at the foot of my bed, go immediately to my refrigerator and take out all the necessaries to build myself a big sandwich. When I was done, I would put all the food back into the fridge, but leave a big mess on the drain board.

Then I would go into the bedroom, eating my sandwich, and take all my dirty clothes out of my suitcase, and throw them on a pile by the foot of the bed, strip off what I was wearing and add that to the pile of laundry. Then I would go into the bathroom and have a nice, long, hot soaker of a shower, making a total mess of the bathroom, too. After my shower, I would dry off, go to my bed, pull it down and have a nap.

After my nap, I would dress and go to the base to check my mail, buy groceries, have a beer with the fellows or whatever, leaving the kitchen, bedroom, bathroom and bed a mess and a pile of dirty clothes on the floor. When I returned from my errands, be it 30 minutes or five hours, my hootch was spotless, as though I hadn't been there at all. Within a few hours all my clothes would be hanging in my closet, freshly cleaned, ready for wearing. My maid must have watched from a secret window from somewhere, for I rarely saw her, but she was always right there, and her work was always perfect.

For this wonderful service, I paid the great sum of $7.50 a month.

About this time our good friend, Buzz Baiz, returned to his

home area of Selma, California, where he had attended his high school tenth reunion. Much to our shock and surprise, he returned with a wife—and her son. Someone had set him up with a blind date for his reunion. He and the girl hit it off famously, ran off to Reno and got married. Buzz brought them home to Udorn. Phyllis and her young son, Keith, fit well into our group of friends.

Dick Koeppe, another resident of the studio apartments, had a little bit of a sadistic streak in him. He would cut a hot dog in half, hollow it out with a penknife, keeping the piece he cut out as a plug. He would fill the center of the hot dog with Tabasco sauce, put the plug in the hole, and feed the spicy morsel to the local stray dogs. They ate it, shook their heads, and came back for more.

One day I bought some decorative bricks to build a bookshelf. In the lattice of one of the bricks I encountered a big scorpion. I captured it and put it in a jar. Not sure what to do with it, I put him outside on my porch, intending to come back and release it into the wild. Unfortunately, in my absence, the sun shifted so that the jar was in the sun, and the poor little bugger died from overheating.

From September 11 to September 25, 1971
the top song on the charts was
"Go Away Little Girl" by Donny Osmond.

30

First Annual Leave – A Trip Around the World

25 September to 7 November 1971

Gary bid his annual vacation to match mine. We both got all of October off. We asked scheduling that our STOs for September and November be attached to either end of our 30-days annual leave. Scheduling honored our requests, so we ended up with six weeks of continuous vacation at full base pay. Upstairs at the company travel office, we bought ourselves tickets on Pan American World Airlines. We wrote in every stop on Pan American's Route One, westbound. This cost us all of $150 each. We were ready for a trip around the world.

Tel Aviv, Israel

Our first stop was Tel Aviv. We had been briefed that we should ask that our Israeli immigration stamps be put on separate papers so that they would not show in our passports. This was to prevent complications should we later desire to visit Arab countries neighboring Israel. (This was only a short time after the "100 Hours War").

We spent the first night at the Tel Aviv Hilton, strangers in a strange land. We checked into the hotel and slid into the lounge for a cocktail. When the cute little waitress came to serve us, Gary asked, "Do you speak English?" Her surprising response was, "Oh, yea, shuah. I'm from Brooklyn." We asked where the action was, took a taxi to the local disco, drank and danced the night away with a bevy of young Israeli girls.

Istanbul, Turkey

We decided Israel was not our cup of tea, so the next day we jetted over to Istanbul. This was my very first Boing 747 ride. I was quite impressed with the large beast of an airplane. We took a riverboat tour of the Bosporus, visited the Blue Mosque and the Topkapi Palace museum, where we saw beautiful bejeweled daggers. We were not terribly impressed with Turkish food.

Out first night there we caught an act in the bar in the basement of the Hilton. An exotic performer sat on a rope swing. As she swayed back and forth she simulated a long, drawn-out orgasm. It was very erotic. We were shocked when we got the bill—drinks were $18 each!

Greece

We popped over to Athens and visited Constitution Square where we enjoyed observing the people from all corners of the world. One afternoon we watched as a young man walked past our table. So intent was he on watching the posterior of the young woman walking in the opposite direction that he walked head first into a lamppost. The local beer was named FIX. For a few days we got our afternoon FIX at Constitution Square. We took the obligatory tours of the Parthenon and the Acropolis. We drank ouzo and danced line dances at the Plaka with the locals.

England

We flew on to London. Neither Gary nor I was a planner when it came to vacation trips. Once we got to London, we really had no clue of what we wanted to do. We rented a Mini Morris, and we bought the ultimate insurance. When I asked the clerk what that meant, he said, "You can bring it back in a basket and you are covered." Good. We might need that.

I am a big guy; Gary was my size. We each had two large suitcases. Somehow we got both of us and all our baggage into the tiny car. We were used to driving on the left side in Thailand and Laos, but the rules in those two countries were more free-for-all than staid London. We encountered frantic traffic, many pedestrians and roundabouts. As I drove, Gary was trying to read a map just to get us out of central London. He would say things like, "There's the Tower of London, turn left," or, "There's Big Ben, turn right." All I could do was take a quick glance and keep driving.

Finally, we got onto the M-5 North, headed to who-knows where? We stopped at a tourist kiosk and picked up some brochures. Then I said, "I remember Hadrian's Wall in the far north, built by the Romans to keep out the fierce Celts. Let's go there." So we did. After about an hour at the endless embankment, we said, "Well, what's next?" Stonehenge. We hopped back into the car and backtracked several hours to see Stonehenge.

We made it a point to go far enough west that we entered Wales. We also drove north far enough to ensure we could say we visited Scotland, adding those two countries to our lists of countries visited.

When we finished touring, I returned the rental car. Turning into the alley (mews) where the rental agency lived, I failed to see a bobby stepping off the curb. I nearly hit him. He glared down at me. I knew I was in trouble, so I rolled down my window and said, "Sir, I am returning this rental car to the office right (pointing) there. If you let me go I promise I will never drive in your country ever again." He waved me on. I have not driven in England since.

We returned to the U.S. and spent some time with family and friends.

We had to be careful about not spending too much time in the U.S. for tax reasons. We did not want to violate the IRS rules and lose our humongous tax break for working overseas. Contrary to what some thought, we did pay U.S. taxes.

A few luggage tags from my scrapbook.

After spending some time with family and friends in California, we flew TWA Flight 745 back to Bangkok the home to Udorn. My first flight after leave was on November 16th.

16 November 1971
H-88
FM Dela Cruz
Copilot not recorded.

Flew one special in the L-11 area (*M-19 on map*).

My first flight after returning from annual leave.

31

Engine Cough with Bob Caron Before a Dangerous Special

19 November 1971
H-88
FM Manalo
I flew copilot for Bob Caron for one special in the L-11 area.

We fly across the Ho Chi Minh Trail

My first special mission after returning from leave we flew out of PS-44 on the eastern edge of the Bolivens Plateau (*O-19 on map*). This high ground overlooks the big valley to the east. The customer assigned us to fly a special across the Kong River to an LZ close to the border of South Vietnam. This entailed passing between Saravane, and Attopeu, both cities held by the enemy.

Immediately after engine start-up, the engine coughed. That was strange, but it caught again instantly and ran smoothly. We did repeated magneto checks and a power check. Everything checked out OK, so we proceeded on the mission. We attributed the cough to a small slug of water in the fuel.

We flew across the Ho Chi Minh trail, found our desired drop off point and headed home. The approach, landing and team insert went without a hitch. We flew almost two hours round-trip mostly over unfriendly territory. I often wonder why we did not receive anti-aircraft fire from the NVA on these missions. Perhaps they did not want to give away the positions of their big guns just to shoot at a couple of old H-34s. Perhaps they thought we were bait to draw them into firing the big guns. During the day we could – and did at times – call in U.S. military bombers to obliterate known gun positions. The enemy knew we had that capability.

We returned to PS-44 and began to shut down the helicopter. When I turned off the electrically-driven fuel pump the engine quit. It was only then that we both realized we had flown that entire mission without an engine-driven fuel pump. The engine had coughed on initial start-up because when I flipped the battery switch from BATT to ON, I deprived the electrically-driven back-up pump of power for an instant, which deprived the engine of fuel pressure for a fraction of a second, causing the cough.

Had that boost pump failed in flight, we would have certainly gone down somewhere on the Ho Chi Minh Trail.

We learned a bit more about our machine that day.

Prostitutes for the Local Troops

There was a time when the press accused the CIA of providing prostitutes to the troops in the field, and, of course, our government vehemently denied it. At the same time, I was in the field delivering the young ladies to the indigenous troops. Perched on the eastern edge of the Bolivens Plateau, a Thai mercenary encampment overlooked the Ho Chi Minh trail. One of the incentives for keeping the troops happy there was for me to bring in a helicopter load of hookers from Pakxe so they could service the troops. I delivered in the morning, they did their duty during the day, and I picked them up and took them back to town at the end of the day. I saw no harm in that.

32

A Heavy Lift Job

23 November 1971
H-81
FM Delacerna

On this date, a customer asked me to go to a valley east of LS-272 to see if I could assist in the moving of a portable saw mill. It sat on a hillside. All the usable timber near it had been logged off, and it was time to relocate the mill.

I landed, shut down, and we looked at the mill. It was in two large pieces, with a base about four-feet wide, maybe 20-feet long, made out of what looked like railroad tracks. The removable carriage above it held a powerful engine that drove the cutting blade to saw logs into lumber. No one knew the weight of it and it had no information placards, so it would do me no good to look at my load charts to see if it was within limits.

My first instinct told me that the base was way too heavy for me to lift. I hooked onto the base with the external load cables and brought in full power. The load would only come slightly off the ground. It was indeed too heavy.

But I had a clear avenue to hover-jump it off the side of the hill. There were no trees in my way, and I had at least 500 feet of clear air beneath me to attain translational lift and carry this beast away. I released it from my external load hook and flew the few miles over to the destination to see where the customer wanted me to place this rig. The delivery point was a clear space beside a small dirt airstrip. I had plenty of room to make a clean approach headed into a stiff breeze. I would be able to maintain translational lift all the way to touchdown of the load.

Picture of an H-34 carrying an external load.
FROM AN U.S. ARMY DASH 10 MANUAL

I returned to the hillside, now lighter on fuel, and made another attempt to lift the rig. This time I had lightened the helicopter by taking out any unnecessary gear and left Delacerna on the hillside. I hover-jumped that sawmill off the ground, flew it down the hillside and delivered it to the nearby airstrip without event.

This was a serious violation of the safety rules, but I knew I could do it. I wanted to help the customer and help the locals accomplish their goal of moving the sawmill, which was important to their economy. I often wondered what might have happened had I gotten into trouble and dropped the sawmill somewhere en route. This was the first of three heavy loads that I carried in a short period of time. Once I was successful with the first one, I felt comfortable in repeatedly breaking the rules about being overloaded.

I found that doing something like this gave me an adrenaline rush that I needed to keep life interesting. PTSD in control once again.

33

BIM Blow Out

1 December 1971
H-59
FM Lorenzo

Flight Mechanic Lorenzo
One of the very best.
PHOTO FROM AUTHOR'S SCRAP BOOK

I was on short final to an up-hill landing on a short dirt strip, somewhere a few miles east of LS-272, carrying a load of supplies to a village. This was a quiet area. No enemy presence or action reported in this area ever. There was absolutely no reason to expect anything out of the ordinary.

About 500 yards from touchdown, both Lorenzo and I heard a loud clattering-hissing noise that grabbed our immediate attention. A picture instantly formed in my mind to match the sound I had heard. I visualized someone holding an AK-47 combat assault rifle at chest level and pushing it away from himself in a forceful horizontal movement. The sound to me was as if that rifle had impacted a cement hangar floor and bounced and slid along on it, the clattering sound reverberating off the hangar walls. It was noise only, no unusual vibrations at all, but something was coming apart!

To this day when I drive I cannot avoid listening to the noises made by whatever vehicle I am driving. Unusual vibrations get my attention right away. Whenever I am in any kind of meeting and someone constantly clicks a ballpoint pen, I have to ask him or her to stop. The intermittent clicking tweaks deeply into my survival instincts.

I completed the approach to the dirt landing strip, landed, and shut down. Lorenzo inspected the H-34, looking for some cause for the noise. At first he found nothing, but then he discovered that the BIM indicator on one of the rotor blades had popped. The BIM indicator told us that the dry nitrogen gas had escaped from the pressurized, hollow, leading-edge spar. The titanium spar of this blade had cracked somewhere. Lorenzo inspected every millimeter of the blade and could find no damage at all. We had taken no hits. I had not chopped anything with my blades … this trip.

H-34 rotor blade hub with BIM indicator popped.
On the H-34s that I flew, the whole indicator was always all white if the blade was healthy. If the pressure escaped from the main spar, what is grey above turned red.
(PICTURE TAKEN BY AUTHOR AT THE ESTRELLA WAR BIRD AIR MUSEUM, PASO ROBLES, CALIFORNIA.)

We thought the problem was simply a release of pressure from the spar and not a serious blade problem. LS-272 was only ten minutes away so we elected to fly the helicopter back to that station, one of the major supply points in our system. As soon as we returned, Lorenzo called in for a new rotor blade. Less than two hours later a new blade arrived by C-123 from Tango. Lorenzo installed it, and we went back to work. Fortunately, it did not need tracking.

(While turning, the blades of a helicopter must rotate in the same plane as they rotate around the rotor head or the imbalance will induce a one-per thump into the system and cause a rough ride. A severe imbalance can cause problems and eventually the vibration will cause damage. Things will begin to come apart).

34

Staying Overnight at Long Tieng

2 December 1971
H-59
FM Lorenzo

The next day, the customer at Long Tieng asked if I would consider remaining overnight. It was an unusual request, but I saw no reason to say no and I was curious as to what I might see. I did learn many years later that a large enemy force had been reported to be in the area and that there was a slim chance that the base might be overrun that night. I was the customers' escape machine. I am sure I could have flown us out of the dark valley at night and south to VTE had there been such a need. I have no idea why there were no Hueys or Twin-Pacs at the customers' disposal, but they must have been tied up on other projects. The S-58 Twin-Pac was an H-34 variant with the nine-cylinder radial engine replaced by twin turbines.

These S-58Ts did special work that took place only at night. I know they used night vision goggles. A friend, who was junior – which made him a perpetual copilot – said he was flying one night when the pilot told him to take off the night vision goggles and look around. Without the goggles, he could not see a thing in the pitch black LZ in which they had just landed. He did not say where it was or what they were doing. Nobody ever did.

That night at Long Tieng, I was called out to execute a rare emergency night medevac. What I saw was mighty gruesome.

Sometimes upcountry we would see the after-effects of the young Thai troops' strong faith in Buddha. Each Thai mercenary wore a chain around his neck with an amulet of his own personal Buddha, which he felt would protect him from harm. I was called out to pick up some poor Thai soldier whose Buddha let him down.

What would happen is this: Thai soldiers would sit around drinking Mekong, the local whiskey. As they got more and more drunk, they would begin to argue about whose Buddha had more protective power. "My Buddha is stronger than your Buddha." The argument would escalate directly in proportion to the amount of booze consumed. Inevitably, one of the troopers would challenge his buddy with the Thai equivalent of, "Prove it!" One of the foolish young Thai troops would then pull the pin on a hand grenade and hold it close to his chest, trusting in his personal Buddha to protect him. I am not aware of any time when the Buddha created a dud grenade or otherwise protected the trooper.

Air America's S-58T Twin Pac, Lao registered XW-PHA.

PICTURE COURTESY OF AIR AMERICA ASSN. TAKEN BY PAUL GREGOIRE.

I was called to haul a bloody mess from the field to the local field hospital. The secure LZ was on a hillside only a half-mile from Long Tieng. As his comrades loaded his body, I could see in the dim light that the soldier was so messed up it was impossible for me to tell which end of the bloody stump had been his head and which end had been his legs. All I saw in the semi-darkness was a bloody mass of palpitating, quivering flesh.

The rest of the night was uneventful.

35

The night medevac at Long Tieng was just a few days before the beginning of

The Battle for Skyline Ridge

18 December 1971 until 4 April 1972

My take on it

The North Vietnamese Army (NVA) had forced the Lao army off several positions controlling the Plain of Jars and was pushing southward to dislodge the Hmong people from Long Tieng Valley. The North Vietnamese Army advanced to the base of Skyline Ridge which overlooked Long Tieng. If they should take this ridge, the way was clear for the North Vietnamese Army to advance all the way to Vientiane. Laos would be lost. Five thousand irregular Hmong soldiers hunkered down in the valley hoping to repulse the North Vietnamese Army force of tens of thousands headed their way. The Hmong soldiers, natives of the area, were determined to hold on to their home. Many of their families and other civilians evacuated to the south.

"The military fight for Laos between the CIA rag-tag army of irregulars under command of Hmong General Vang Pao and two invading North Vietnamese Divisions under command of PAVN (People's Army of Vietnam) General Nguyen Huu An came down to a single ridgeline."

– *Customer, James "Mule" Parker.*

As it pushed south, the PAVN army made two huge mistakes after taking the Plain of Jars. They were equipped and trained for plains warfare and had never trained for mountain warfare. They also were poorly supplied. Once the NVA soldiers tried to climb the ridges and hills surrounding Long Tieng, they suffered greatly for lack of food, ammunition and especially, water.

The NVA attacked the numerous outposts (Charlie sites) on Skyline Ridge and managed to temporarily take over one or two, but they could not hold the hilltop redoubts. These outposts were held by Thai mercenary troops who did well for themselves. The NVA actually got two tanks over the ridge, which should have been disastrous for the locals, but one CIA case officer had the foresight to place a pair of anti-tank mines on the narrow dirt road to Long Tieng. Both the tanks were blown up, not only stopping the immediate threat, but also blocking the narrow, ledge-like dirt road from further traffic.

The final defeat for the NVA for this campaign came when observers reported them amassed in a place known as General Vang Pao's farm. The location was quite close to Skyline Ridge, but the case officers took a risk and called in a U.S. Air Force B-52 strike (called an "Arc Light"). Hundreds of 250-pound bombs blew the entire PAVN regiment into smithereens. The PAVN general, Nguyen Huu An, admitted defeat and retired his troops back towards North Vietnam.

For ultimate detail about this battle, the reader is referred to the below list of books by former customer James Parker, aka "Mule." As a customer, he had a much better overall view of what happened on the ground during this period of time at this place.

1. *Timeline: Battle for Skyline Ridge*

2. *CODENAME MULE*

3. *Covert Ops.*

4. *The Vietnam War Its Ownself.* In my opinion, this is the best of his works, and it goes into more detail about the battle.

I worked the Long Tieng 16 days during this period, but I was not directly involved in the actual battle. I know I carried supplies, including tons of "hard rice" (munitions) to the troops on the various "Charlie" sites on Skyline Ridge.

An Air America H-34 landing on Skyline Ridge between sites CA and CT.
PHOTO BY AUTHOR FROM HIS PERSONAL SCRAPBOOK.

Later, as a Bell pilot, I worked these sites for many hours a day, day after day. Some days flying almost 12 hours.

The NVA had 130mm howitzers that fired directly into the valley from the north. They usually fired only at night so that the big guns could not be spotted by observer aircraft. Rare stray rounds landed during the day, usually on the hillsides south of the airstrip. Of course, that means the rounds passed thru the airspace above the airport.

There were rumors about the possibility that an enemy Mig jet fighter might try strafing the valley. We didn't give this much credence. We knew the U.S. military had air superiority of the area. To do so would be a suicide mission for a Mig pilot.

February 1972 began the onset of another smoky season. For weeks, visibility dropped down to hundreds of yards in most places, making navigating a challenge. One trip, I worked the area in the middle of Laos that wraps around the northeast corner of Thailand (*I-12 on map*). I had never worked this area before and navigation was most difficult because I did not know my way around and I could not see any landmarks. I had to fly around at about 80 knots. I also always flew with my landing lights on so that I was more visible to any other aircraft that might be working in my area. As much as I could, I followed roads.

Long Tieng airport, looking west.

PICTURE COURTESY OF PETER AND BAYSONG WHITTLESEY, AUTHORS OF *SINXAY, RENAISSANCE OF A LAO-THAI EPIC HERO.*

Air America operated an entire fleet of aircraft out of the "Secret CIA Airbase at Long Tieng." At one time, this was one of the busiest airports in the world. Hundreds of flight operations every day supported the Lao army in its battle to hold off the North Vietnamese Army intent on taking over Laos.

Secret CIA air base Long Tieng from afar with Skyline Ridge to the north of it. Looking northwest.

PICTURE COURTESY OF PETER AND BAYSONG WHITTLESEY

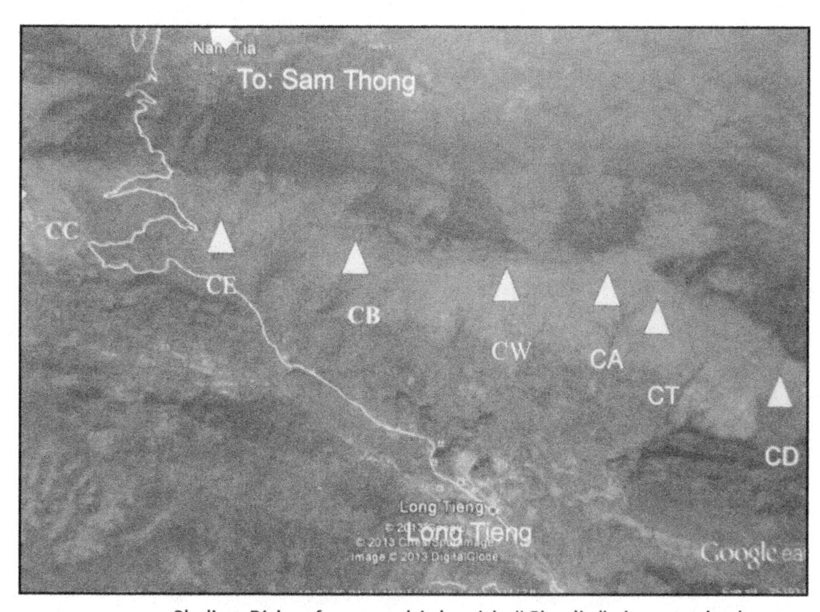

Skyline Ridge from on-high with "Charlie" sites marked.
The "Charlie" sites are where the Thai mercenaries heroically held off the NVA.

PICTURE BORROWED FROM PARKER'S "BATTLE FOR SKYLINE RIDGE."

A collection of Air America Hueys on the Long Tieng ramp, safe behind the karst.

PHOTO COURTESY OF AIR AMERICA ASSOCIATION.

36

Another Heavy Lift Job

20 December 1971
H-89
FM Delacruz
Copilot Wilbur

The customer at Pakxe (*M-19 on map*) asked me to lift the base for a 155mm howitzer from a small hilltop east of the airport and deliver it back to the airport for repairs. I said I would go take a look at it. This was much the same situation as the portable sawmill I had carried in late November. Had the various customers talked among themselves and learned who to ask to take on such a dangerous job? The gun carriage was too heavy to lift under normal power, but I repeated my procedure as before. Again I flew around until I was low on fuel, emptied all the excess gear, and removed my copilot and the FM from the helicopter. I hover-jumped the heavy load off the hill into a stiff breeze. It was no strain to carry it to Pakxe airport without incident. I had a stiff head wind at the airport where I dropped the piece, so I was able to use translational lift to my benefit. I returned to the hill, retrieved FM Delacruz and copilot Wilbur and called it a day.

PTSD adrenaline rush for the day achieved.

Pakxe, Laos.

Ms. Ott's Engagement Ring

Of course, after a hard day's work, every young, hormone-driven mercenary aviator needs a little diversion. Most of the bases had a small American community and there were third-rate movies and card games, but mostly we went out to the local bars and chased the local girls. After all, third-rate movies and card games are poor fare for young mercenary pilots in search of love. Chase isn't really a good word, because none of the girls were difficult to catch. All the towns had little bars that catered to the Air America pilots and crews and each bar had working girls. We availed ourselves of their services when the need arose. That was just part of the experience.

One amusing incident happened in Savannakhet at the nearby bar named "The Blue Fox," after the infamous bar by the same name in Tijuana, Mexico. (I visited the original Blue Fox when just out of high school.) One of the bar girls, Ott, was a rarity among Asian women. She was a treasure to us American men, who liked our women busty. She had huge boobs. One night, Ott suddenly became less friendly and said that she was no longer available. She had fallen in love with one of the helicopter mechanics and she was engaged to marry him. She proudly displayed the ring the mechanic had given her. To her it was the Hope Diamond.

The ring looked a little dingy and it had no stone of any sort. It was a plain black, round wire-shaped band. We asked her what kind of ring it was, and she replied proudly, "Velly special, is lubbah o'ling." It took us a while to figure out what she was saying in her poor English. What she had on her finger was the rubber seal from the fuel system of a helicopter, a rubber O-ring.

We never told her.

21 December 1971
H-89
FM Delacruz

Copilot for "Pappy" Wright. We fly one special mission near PS-44.

23 December 1971
H-89
FM Wade

I fly copilot for Gary. We flew one special mission also near PS-44.

25 December 1971 – Christmas Day.
H-89
FM Punzalan

I fly copilot for Gary; we fly together on two special missions near Pakxe. Then we departed in separate H-34s. The Company made it a point to try and have all the married men at home on this special day, so all the single men worked. We did not mind. After the specials, we were within radio shout of each other, so we were able to talk a bit over the air as we worked the area west of Pakxe on the Bolivens Plateau. I flew out of LS-155 (*N-19 on map*).

27 December 1971

Capt. George L. Ritter, First Officer Roy F. Townley and crewmen Edward J. Weissenback and Khamphonh Saysongkham are shot down and listed as MIA, near LS-69, close to the China Road.

27 December 1971
H-54,
FM Lorenzo

Tom Grady is my copilot. We fly five specials in PS-1 area.

Follow up.

After Air America, Tom Grady was flying a Huey 205A-1 for Arizona Helicopters on contract for the U.S. Forest Service in southern California. The USFS was experimenting with using night vision goggles to fight fire at night, when the fires are usually quieter. Tom collided with another helicopter and died instantly.

When I heard about this, I was told that Tom was sitting on the ground refueling when the other helicopter landed on top of him.

A community college fire-training center in Southern California is named after Tom.

Here is the incident report:

"Incident Name: Middle Fire, Angeles National Forest

Date: July 24, 1977, 2230 hours

Personnel: Thomas Grady

Age: 32

Agency/Organization: pilot for LA County Fire, under contract with the USFS, flying a Bell 205 helicopter owned by Arizona Helicopters

Position: pilot

Summary: On the night of July 24, 1977, a Los Angeles County Fire helicopter (LACO 14, a Bell 205A-1) and the USFS Rose Valley helicopter (H-29, A Bell 212) collided mid-air while inbound to the Mill Creek Heliport on the Angeles National Forest. At the time of the crash, both helicopters were involved in night firefighting activities, dropping water on the Middle Fire and were under contract with the United States Forest Service. Pilots of both helicopters were operating with Night Vision Goggles (NVG). Radio frequencies were saturated, and the pilots had trouble communicating with the helibase manager. The county helicopter (a Bell 205A-1) was piloted by Thomas Grady and Theodore Hellmers, both employees of the Los Angeles County Fire Department. The helicopters collided while being maneuvered to land. Grady was killed in the collision and Hellmers suffered severe injuries. (Litigation went on for years.) The two pilots of the Forest Service helicopter (a Bell 212) were also injured but not so severely. The Middle Fire was 3800 acres."

From: http://wlfalwaysremember.org/incident-lists/237-thomas-grady.html

37

Our Base at Ban Houie Xai

When I worked out of Ban Houie Xay, (*B-8 on map*) most of the work was routine. Once again, I resupplied Lao or Thai troops on nearby hilltop outposts. It seemed to be a quiet area and I remember having no difficulties. At night, we parked the H-34 in the soccer field of a local school. Since we usually landed just before sundown and departed at first light every morning, our machines were never a problem for the school kids.

At the Ban Houie Xay company hostel, the water in the shower was always slightly oily. Finally, someone realized that the well providing water for the old French chateau was downhill from the diesel fuel storage tank that supplied fuel for the hostel generator. Every time anyone filled the fuel tank a little diesel spilled. That spillage eventually seeped into our well, thence into our bathing water. I was tempted to place a sign in the bathroom:

DANGER - No Smoking in the Shower

It was in this area that my friend Gary discovered the Kao drums. These were apparently ancient bronze drums about 18-inches high and about 24-inches in diameter. The walls of the drums were very thin and delicate. Some had ornate decorations on them formed of the same bronze. These intricate works of art had been hand-forged in a single piece by primitive tribes people centuries before the French came to Indochina. It seemed impossible that these primitive people could forge something so delicate and intricate. The technique for creating these is lost to antiquity. Some said that they were used by local tribes in fertility rituals.

Ancient Kao drum with a ruler for scale. PHOTO BY AUTHOR.

We all scoffed at Gary for buying two of these things at the exorbitant price of almost $200 each. He said he wanted them to use as end tables for his future bachelor pad. We considered him slightly nuts. Not long after he returned to the U.S. he had them appraised at close to $5,000 for the pair. After he returned to the U.S. it became illegal to export such artifacts out of Laos. They were considered Lao national treasures.

I have no idea whatever happened to Gary's Kao drums after he was killed in the helicopter crash in 1975. I believe whoever packed up his belongings helped himself to the drums and a few other artifacts. Some other things disappeared, too.

Another time at Ban Houay Xay, I had a long conversation with Al Schwartz of Continental Air Services (CASI). Al flew the Pilatus Porter for CASI. He was a talker, and his mind wandered a lot, so once he started a story, he veered off on a dozen tangents before finally returning to the end of the story. One night he rambled on for 20 minutes, but finally finished his story about flying F-4U Corsairs for the Marines in Korea.

The aircraft Al Schwartz flew in Korea.
F-4U Corsair (FROM WIKIPEDIA)

Al finally finished his story with, "On a bombing run, I was a bit low when I released my bombs. There was a loud BOOM, and my engine quit. I looked down between my feet and could see the ground. I blew myself out of the sky. I belly-landed behind enemy lines, and the Red Chinese almost got to me before I was rescued by the Marine Corps infantry unit that I was supporting."

38

Life in Thailand

We were not allowed to have personal weapons in Thailand. We all had "survival" weapons at work, but we left them in our lockers on the base, or checked them out from the company armory before each trip. We never took them home. Bob Sweeney's house guard accused Bob of having a gun at home. Bob was cited by the local police, and had to go before a Thai judge to defend himself. None of our people could speak Thai well enough to protect himself in Thai court, so Bob was forced to hire a local Thai lawyer.

Came trial day Bob's defense lawyer asked the guard on the witness stand what time of day that he saw Bob with the weapon. The guard said, "Two o'clock." The lawyer asked the guard how he knew the time that he saw Bob with the firearm. The guard said he looked at a clock. The lawyer then asked the guard to look at the clock in the court and tell the judge the time. He could not. He did not know how to read a clock. The case was thrown out, but still it cost Bob $1,000 in lawyer fees.

Bob died in a helicopter incident in Mexico about 1983 along with our former Air America instructor Danny Carson. Two great guys gone.

Life was cheap in Asia. The Thais were adamantly anti-communist. They had a serious problem with gunrunners bringing illegal arms into the country for communist subversives trying to overthrow the Thai government. A gang of gunrunners was captured by the Thai police near the Mekong River, across from Vientiane. The evil fellows were tried by a military court, found guilty, and sentenced to death by firing squad.

It was difficult to get a firing squad together to shoot someone because most Thais are Buddhists. As Buddhists, they tend to be very gentle, loving people and try, in the Buddhist manner, to not kill any living thing.

The authorities came up with a very creative solution. Each condemned man was tied to a post. A grid of four outer posts was placed around him. The four posts were then wrapped by cloth, obscuring the condemned man from sight. A bull's eye painted on the cloth, became the target instead of a person. All the firing squad had to do was shoot at the bull's eye and, in effect, not shoot at the person inside.

The last paragraph of the *Bangkok Post* article, where it cited the difference between a real execution, and a Thai style execution, said, "... and several M-1 carbines on full automatic emptied full clips into the bull's eye."

Thailand had no communist insurgency problem after that.

39

Christmas Fireworks

Just after Christmas 1971

In Udorn with a few days off, Gary and I ran into a fellow H-34 pilot and former U.S. Navy pilot, Sandy Sandt. He invited us over to his house for some Christmas cheer. We had a few drinks at his home bar.

Gary Connolly and Sandy Sandt in heated discussion at Sandy's bar, just after Christmas 1971.

Sandy was very proud to show us his Christmas present – a pellet pistol his wife had bought him. He needed the pistol to chase away the stray dogs whose barking often kept him awake at night. The idea was that he would go outside and shoot the dogs with the pellet gun and they would go bark elsewhere. Had we been allowed to have real weapons in our homes, there might have been a large number of dead dogs around town.

Sandy was very proud of his new pistol; he wanted to demonstrate it. He said, "Watch this." He cocked it, took aim at the family Christmas tree at the other end of the living room, about 20 feet away, and expertly shot the pretty glass star off the top of the tree, shattering it to bits.

His wife was not pleased.

30-31 December 1971
H-15
FM Wade/Ramos
My copilot 12/30 is Richie; I am copilot for Kawalek on 12/31/71

Two Days of Crane Chase

At this time the battle for Skyline Ridge was coming to a head. The North Vietnamese Army was about to overrun the base at Long Tieng. The Company was retreating 20 miles to the south to LS-272 and building a whole new base there. It was a busy, frantic time. U.S. Army CH-54 Air Cranes were called in from Vietnam to help move massive amounts of material in the building and supplying of the new base. We flew escort for the Army Flying Cranes to guide them and to rescue their crews should they crash. One Crane did later crash near LS-272. Because of that crash, I soon was called upon to do another heavy lift job.

A routine rescue near Vientiane.

A Lao army H-34 had an engine failure and made an uneventful landing a few miles north of Vientiane. I picked up the crew and delivered them back to the capitol.

6-7-8 January 1972
H-73
FM Ranny Lacson
I fly copilot for Gary.

We fly seven special missions these three days, near Pakxe. (*M-19 on map*).

SAR for Royal Lao Army H-34 Near Pakxe

Contrary to our USMC experience in Vietnam, we Air America pilots did not have to go into hot LZs to pick up wounded Lao soldiers or to do emergency resupply missions. The Lao army had its own H-34s and pilots who did such perilous things. The Lao crews were very good. Occasionally, we flew Search and Rescue (SAR) backup for their emergency missions.

On one mission, the customer briefed us to rescue the crew of a Lao helicopter crew that had been shot down while landing in a hot LZ southeast of Pakxe. We got all the information from the customer, appointed crews by seniority, and carefully briefed the appropriate procedure to accomplish this mission as safely as possible. By the time we got on scene, one of our own, Hal Miller flying single pilot, had already slipped in low and fast, and successfully accomplished the rescue without backup. It seemed that sometimes Hal had more balls than sense. Hal did a similar thing a few months later.

On a team extraction somewhere north of the Luang Phrabang area, the lead helicopter got hosed with bullets. Their engine quit and they autorotated onto a hilltop, landing in tall elephant grass. The crew evacuated the machine and immediately became scattered in the tall grass. They were all in a severe panic as they had just been shot down and knew the enemy was nearby, searching for them. They had no way to communicate with each other. They did not want to holler for their crewmates for fear of attracting the enemy. None of them knew when his next step would be right in front of an NVA soldier with an AK-47 combat assault rifle.

One of the pilots used his emergency radio to call the Search and Rescue helicopter orbiting overhead. The captain of that H-34 SAR refused to descend, fearing that to do so would only get him shot down beside his leader. A good friend of mine was the copilot on this mission, but he could not convince that pilot in command to descend to aid the stranded crew. Things were looking grim for the crew on the ground.

Hal Miller, working about 20 miles south, single ship, single pilot heard the Mayday calls of the downed crew. He immediately pulled full power on his engine and headed in their direction, even though he was low on fuel. He managed to slip under the SAR helicopter, pull all three members of the downed crew from the tall grass and egress the area without taking a single hit. Returning to a safe area for fuel, he had not even fumes left in his fuel tanks. We never heard whatever happened to the indiginous team on the ground awaiting extraction.

10 January 1972

Filipino Flight Mechanic Ernesto Cruz was killed when an explosion occurred while he was offloading cargo south of Sam Thong (*E-10 on map*). He was a nice fellow and had crewed for me a few times.

19 January 1972
H-87
FM Wade

I fly copilot for Dave Kendall on five specials near Ban Xieng Lom.

From January 22nd until February 5th,
The top song on the charts was
American Pie.
by Don McLean

Twelve Specials near Louang Phrabang

30 January 1972
H-87
FM Ernie Cortez

My copilot is John Ferris.

Flying out of Louang Phrabang with John Ferris as my co-pilot, we made 12 special landings in one hour. That boosted my paycheck $600 for that month for those 60 minutes work. We simply ferried troops to the top of a nearby hill that was supposedly contested. We returned into the valley to load some more. I swear, and John agrees, that by the time we took the last load up, some of the troops were the same ones we hauled up on the first load.

Author, mid-1972.

25 JANUARY 1972

PRESIDENT NIXON ANNOUNCES A PROPOSED EIGHT-POINT PEACE PLAN FOR VIETNAM. HE REVEALS THAT SECRETARY OF STATE, KISSINGER HAS BEEN NEGOTIATING IN SECRET WITH NORTH VIETNAM. HANOI REJECTS NIXON'S PEACE OVERTURE.

Around this time, I was told to take a cargo and a few passengers to a site that was far to the northwest. The strip was above 8,000 feet in elevation. I planned my fuel carefully so as to have enough to return to Long Tieng, but to be as light as possible for landing at that high elevation. Fortunately, there was a good landing strip. I was able to make a roll-on landing and deliver my load without incident. Normally the highest any H-34s worked was 4,000 feet in the hills around Long Tieng.

We often worked the area around Louang Phrabang, the Royal Capital of Laos (*E-8 on map*). We *farongs* pronounced it, "LONG Prah-BONG," or simply called it "LPB." The small town in northwest Laos was hardly more than a large, dusty village perched on the bank of the Mekong River. To my knowledge at the time, enemy action in this area was rare, but it was only a short distance from the Plane of Jars where there was plenty of action.

While working this area we stayed in an old French two-story hotel the company had converted into a hostel. It was in the shape of an "L," surrounded by tall teak trees. A thatch-roofed gazebo bar sat in the tiled patio of the L. A profusion of lush jungle plants grew right up to the edges of the hotel borders. Rumor was that a small Asian bear lived in a cage just behind the hostel, but I never bothered to verify this.

Our East Indian hostel manager, Abdul, preferred to cook curry dishes instead of American style food. After a hard day's work flying around the area, we would have delicious curry dinner at Abdul's restaurant and rendezvous at the bar for a few cocktails. Before the twelve-hour limit, we would retire to our rooms for a night's rest and perhaps to drink a bit more in private.

There was an old woman, whom few of us ever actually saw, who performed a service for every one of us at one time or another. She was the local French-trained oral prostitute. At night, after everyone had turned in, she would come to each room and try the doorknob. If the door was unlocked, it was an invitation for her to enter...the only invitation she needed. She would quietly and carefully creep into the room, find the pilot or crewman in his bed and perform oral sex upon that awaiting man. She would then quietly slip out and continue on to the next waiting customer. We paid her generously.

Near this exotic small city in remote Laos called 'Long Pra Bong,' I would soon have one of the greatest adventures of my entire life.

40

Worry Four

The Rescue of Raven 1-1

"If you are in trouble anywhere in the world, an airplane can fly over and drop flowers, but a helicopter can land and save your life."

–Igor Sikorsky

As I continued flying south toward Udorn, my rotor blades flogging the air into submission, my imagination revved up into turbo-overboost trying to figure out why I had been recalled. Temperatures and pressures in my mental anxiety gauges were approaching the tops of their yellow arcs.

Yet another reason crossed my mind for being in "hac." I had recently ruined an $85,000 engine in one of our helicopters. True, it had been during the daring high-hover, hoist rescue of two downed Air Force personnel who had crashed after being shot full of holes over the Plane of Jars, but could that be a reason? The company had been getting a lot of grief about aircraft losses.

31 January 1972
H-87,
FM Ernie Cortez
Copilot Dick Koeppe

Koeppe and I were working the Louang Phrabang area single ship. We had flown about five hours resupplying the local troops at various hilltop outposts up the Nam Ou River, a tributary that entered the Mekong five miles north of LPB. We were returning to LPB at the end of the day, looking forward to a cool one at the Gazebo Bar. We had flown past a huge Buddhist temple, with caves cut into the cliffs at the confluence of the Nam Ou and the Mekong. I badly wanted to fly down close to the temple and take a close look at it, but two things stopped me. First, I felt it would be plain rude to interrupt the meditations of the monks with our engine noise. Secondly, from my Vietnam experience, I had an aversion to flying too low and slow over any place where there might be enemy. I had an allergy to bullets.

As we approached the airport, we heard a cacophony of radio traffic. This was unusual because by this time all aircraft should be on the ground and headed to the bar. As we got closer, we observed several aircraft buzzing around the airfield in a clockwise pattern. When the radio transmissions became clearer, we realized that everybody was trying to locate a Raven spotter pilot who was down somewhere nearby. He had been shot up over the PDJ and crashed while limping home.

With brass balls larger than grapefruit these Raven pilots flew low and slow over the Ho Chi Minh trail, looking for enemy fuel depots, truck parks or supply dumps. When they spotted something, they could call in bombers from several different sources and destroy the enemy supplies. The enemy knew what the Ravens were doing and did not shoot at them because they knew to do so was to have hell rain down from the sky. It was a cat and mouse game. If the spotter plane made one circle and moved on, the enemy knew they were safe. If he continued to circle, they knew they had been seen, and then they would shoot at the Ravens. Of the approximately 128 Raven pilots in the war, nearly a third of them died.

We joined the other aircraft circling the airport. Nothing concrete seemed to be happening. The main problem was that so many pilots were talking on radios at the same time that no one could have good communication with the downed man. All the other pilots kept "stepping" on each other's transmissions. That also precluded getting a good ADF (automatic direction finder) fix on the downed aircraft. That simple instrument, if not blocked by all the chatter, would have instantly given any pilot the proper direction to fly.

A Raven 0-1 Bird Dog spotter plane.

On our first orbit, I made a small observation: although the downed Raven's radio transmissions were very garbled, there was one point where he came in loud and clear. In that moment of clarity, I heard the phrase, "Up a canyon." That made sense to me. His being up a canyon would explain the variations in radio reception. Radio waves travel only in a straight line and cannot go around corners or through solid objects like mountains. I heard him clearly again at a second point of our orbit, opposite from where I heard the first clear message. In my mind, I connected those two points in a straight line. I turned and followed that line eastwardly, towards the Plain of Jars. There were mountains with a canyon between them.

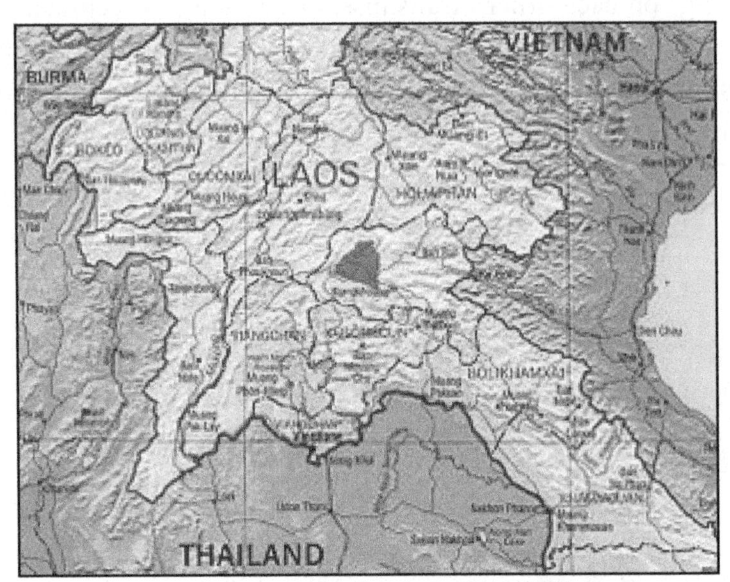

Plane of Jars, central Laos. (Dark triangle) (WIKIPEDIA)

We flew only a few miles when I heard the stranded pilot say, "Helicopter, you are right overhead!" I did a quick stop and hovered there. You can do that in a helicopter. Filipino FM Ernie Cortez spotted the men on the ground. About this time, the noise on the radio from the other aircraft became overwhelming and bothersome. I told copilot Dick to tell all the other aircraft to get off the air and "maintain radio silence, rescue in progress." As Dick was doing that, Ernie was letting down the hoist cable. I focused on a nearby treetop and established a stable hover above 150-foot tall teak trees on the steep hill side. This would be a challenging high-hover hoist recovery.

We hovered there for nearly 15 minutes, all the time near maximum rated power and getting heavier all the time as we hoisted the first, then the second airman, from the downed aircraft to safety. All the while Dick monitored the radios and monitoring the engine instruments. Ernie did great job with the hoist. Finally, we had both of the crewmembers aboard. We hopped the short flight to LPB and unloaded the hitchhikers.

Mike Kelly's crashed 0-1 Bird Dog. PHOTOS COURTESY OF MIKE KELLY

U.S. Air Force Captain Mike Kelly, arm broken from his crash, with two local T-28 pilots.

I was a bit disappointed after this rescue. Rumor had it that when a helicopter pilot rescued another combat pilot, the rescuee is supposed to give his pistol to the rescuing pilot. The saved pilot can always claim it was lost in the crash and it will be replaced, no questions. He is also supposed to buy his rescue pilot a bottle of booze.

The only injury between the two of them was that the pilot had a mildly broken arm. The Raven waved a feeble "thank you" with his good arm as the ambulance crew swooped him up and took him away before I could even shut my helicopter down. I didn't even learn his name.

At this point, I began to realize that I might be in a bit of trouble. The maximum time for takeoff power on my engine was limited to five minutes. I had just put 15 minutes of near maximum power. In effect, even though it was still running well, it might be ruined – $85,000 trashed. It would require a major inspection, if not a complete overhaul. Would I be in trouble, would I get fired? I called Udorn and told the chief pilot. He asked me if I felt Okay ferrying the aircraft home. I said "yes." On the way home I wondered if I would again be in "hack" for ruining the engine, but I felt good about the rescue.

I returned the aircraft to Udorn line maintenance, did the appropriate paperwork and went into operations. The only thing anyone said, "Good job, Bill," came from Chief Helicopter Pilot Wayne Knight. The next day I got another aircraft and went back to work.

Could the ruined engine be the reason I was being recalled?

In flight school and throughout my military flying career, it was ingrained into my psyche that a good pilot never leaves a damaged and potentially dangerous aircraft for the next pilot to fly. How would you feel if you overstressed an aircraft in some way, and your good buddy died in that same aircraft the next day because you did not write up a serious deficiency? Better to report your dumb mistakes and be a bit embarrassed than kill a friend—or a stranger.

Follow up:

Thirty-four years after I departed Udorn, I attended the funeral of "customer" Tony Poe in Sonoma, California, which just happens to be my hometown. I had flown for him in northwest Laos at LS-118A,

Nam Lieu (*E-6 on map*).

I stood on a street beside the Saint Francis Solano Catholic Church on Napa Street, awaiting my good friend Dennis Kawalek, who was driving over from Santa Rosa. A fellow approached me, took a look at my Air America baseball cap, and said, "You were with Air America, I see." Yes, I told him, and introduced myself to former U.S. Air Force pilot Gene Hamner from Lodi, California. Dennis showed up, and the three of us attended Tony's mass. Afterwards, we joined the mourners at Tony's house east of Sonoma for a celebration of Tony's life.

In talking with Gene I found out that he too, had been a Raven spotter pilot, Raven 1-2, working out of Long Tieng. I asked him if he knew who got shot down at Louang Phrabang on 31 January 1972. He said he did not know, but he would find out. A few days later he gave me the address of Raven 1-1, whom I had rescued. I wrote a letter to that fellow, Mike Kelly, and sent him the copy of my version of the story that I had previously written down. At the bottom of the letter I post-scripted, "By the way, I think you owe me a bottle of booze." About two weeks later I got a hand-written letter back from Mike, saying, yes, he thought I was the Air America helicopter pilot who had rescued him. His postscript said simply, "P.S. The check is in the mail."

About a week later UPS delivered a small package that contained a bottle of Johnny Walker Blue Label. I had never heard of Blue Label before.

In Laos, I drank JW Black when I could get it and Red when I could not get Black. For a while after my Pitsanoulk experience, I became a convert to Courvoisier Cognac, but Blue was a new thing to me. I happened to be near a liquor store a few days later, so I decided to see what a bottle of Blue might sell for. $200! Mike Kelly had sent me a $200 bottle of Scotch. I emailed him in Texas and told him that I did not drink much anymore, and he should come out to California and help me drink that bottle of Blue. He agreed to do so. He was a senior captain with Continental Airlines, so he could travel much easier than I could. We set a date to meet.

Johnny Walker Blue Label.
(PICTURE FROM WIKIPEDIA)

"Better a free bottle in front of me,
Than a pre-frontal lobotomy" (ANONYMOUS)

I told my older daughter, April, about the upcoming event. She and her husband, Daniel, were independent filmmakers in San Francisco. She insisted, "Dad, we have to get this reunion on video." I liked the idea. I met with the filmmakers at their home one day and they shot several hours of video. I not only talked about the rescue of Raven 1-1, but I also was able to record a lot of flying stories about my experiences both in the war and afterwards. These are archived for my grandkids. April also insisted that Mike and I not meet until she had a chance to interview him separately. That way, we would not cross-contaminate each other's stories.

When Mike arrived at San Francisco airport, April picked him up and brought him to her beautiful condo in San Francisco. Gene Hamner, Dennis Kawalek, my wife, Carla and I, had to make ourselves scarce for a few hours while April and Daniel interviewed Mike Kelly. When they were all finished, we returned to the apartment.

I met the Air Force pilot that I had rescued 34 years before.

We greeted, hugged, and got down to getting to know each other. We opened the bottle of Blue and started drinking and toasting. We drank not only the Blue, but also several other bottles of liquor and wine and champagne. Between 2 p.m. and 3 a.m. the next morning, we drank at least five or six bottles. We ordered in Chinese food—a lot of Chinese food. The four of us toasted everyone and everything and told a million war stories. I made a toast to Dick Koeppe, who had been my copilot on the rescue. I toasted him especially because, only a few months before this meeting, Dick had "lost the battle with PTSD" (Post Traumatic Stress Disorder). I am sad that Dick was not there to meet Mike and Gene and to share the fine whiskey.

Dennis Kawalek, Mike Kelly, Bill Collier, and Gene Hamner toasting Dick Koeppe.
PHOTO BY APRIL DAVILA.

An interesting side point here is that Dick and I were hired on the same day. He was older than I. He should have been senior to me, but some mix-up in the company paperwork made him junior. He complained about the mix-up several times. Each time I told him, "Dick, go to admin, tell them you are older than me and that we were hired on the same day, and that you should be senior to me. I will agree with any change they make." He said he would, but never did. Had he, this might have been his rescue instead of mine.

April and Daniel got every minute of that 13-hour gathering on videotape. I took that footage and a documentary that I have about Marine Corps H-34 flying in Vietnam called "Helicopter Heroes" from the History Channel, and several more I have about Air America, to Ms. Michelle Stikitch in San Francisco. She edited it down to one 25-minute documentary. It can be viewed on YouTube: "The Rescue of Raven 1-1."

Watching that video a few weeks later, I learned two things about this event that I did not know before. First, the area where I picked up Mike and his back-seater was HOT! The enemy was close on Mike's trail and was trying to capture or kill him. If they had gotten a bit closer, we would have taken fire, and probably hits, hovering there. The potential for getting shot down was high.

The second thing I learned watching the video is that we did not have Mike aboard the helicopter when I moved off from my high hover. I remember hearing a voice saying, "Let's get out of here!" I thought it was crew chief Ernie Cortez, so I departed my hover. Unknown to me, Mike was still dangling about 20 feet beneath the helicopter, and I dragged him through the tree- tops. It was Mike who said, "Let's get out of here," not Ernie. Mike chose to be dragged through the treetops rather than (literally!) hang around and risk getting shot down a second time.

Ignorance is bliss. I thought at the time this area was cool, and I thought Mike was inside the helicopter when I moved off. Being dragged through the trees is not what broke Mike's arm. It was broken in the crash.

More follow up:

Because of his perilous mission as a Bird Dog pilot in Asia, Mike was given his choice of assignments upon his return to regular service. He chose to fly U-2s for the rest of his Air Force career. Mike retired from the Air Force and had a second career as a captain with Continental Airlines. He lives in Montana and goes fishing a lot.

Gene Hamner retired after a career as a crop duster in the

Stockton, California area. Before his career with the U.S. Air Force, Gene was a U.S. Forest Service smoke jumper. We three continue to communicate.

Dennis Kawalek has been a life-long friend since Navy flight school. He married his Thai girlfriend, Dang, and they settled in Santa Rosa, California. After the Asian refugees came to the United States, many settled in our area. He and Dang started and managed an Asian grocery store, *Siam West*. They had three beautiful children. Sadly, his wife died when the children were young. Even sadder, his older daughter died in her sleep in her early twenties. His younger daughter and son are doing well in life. The daughter recently (Mid-2019) made Dennis a grandpa for the forth time. We keep in touch.

Author's note regarding the rescue of U.S. Air Force pilots:

It was not exceptional for one of us to be in the area when an Air Force pilot had to eject after sustaining damage over North Vietnam or Laos. Many times we would swoop in, gather up the crew, and have them back to their base before the Air Force could get permission from D.C. to cross the border from Thailand.

From what I heard, this is an approximation of what went on in the Air Force rescue regime. I am sure it is one sided, exaggerated and prejudiced, but it makes a good story.

When an Air Force pilot started calling Mayday or ejected out of his jet over Laos, the Air Force would launch two CH-53 "Jolly Green" helicopters from Nakon Phenom, northwestern Thailand. Since they could not cross the border into Laos without clearance from the Pentagon, they would start an orbit at the border. Before they could get clearance, they realized that they did not have enough fuel to continue on the mission and make it back home. Then the Air Force would launch a second pair of CH-53s as back up. Because there were now four aircraft in the sky on the same day, they required fighter escort, the Air Force then launched a section of fighters (two, maybe four).

It was not in the Air Force mentality to have that many aircraft in the sky by themselves. They needed some kind of airborne control to coordinate this many aircraft in the same sky on the same day. They had to have CAP overhead to command the overall situation. Before any aircraft could cross the border, the USAF might have eight or ten aircraft milling around at the border, awaiting clearance to enter Laos.

Meanwhile, a solo pilot in an unaccompanied Air America H-34 in the area would hear the distress call on guard (universal emergency) channel and the pilot would cruise over and pick up the pilot and take him home. The USAF would then cancel its massive response. That happened many times.

One of our senior pilots, Charlie Weitz, was sitting in the Udorn U.S. Air Force Base officers' club one night, having drinks with some Air Force pilots. They all introduced themselves and compared notes about what they did for a living. One of the last to speak was a USAF CH-53 Jolly Green pilot. He was all puffed up with his own importance as a Jolly Green rescue pilot. When he was finished blowing about how great he was, he asked Charlie what he did for the war effort.

"Oh," Charlie replied, "I do your job."

(Charlie Weitz died late October 2016).

4 February 1972,

Crewman Khamouth Sousadalay is killed when his C-7A Caribou crashed on landing.

7 February 1972
H-83
FM Lorenzo, copilot Chuck Frady

We fly one special mission in the Louang Phrabang area.

11 February 1972
H-80
FM Lacson

Jess Hagerman is my copilot. We fly two special missions in the Pakxe area.

26 February 1972
H-88
FM Punzalan

We Rescue a Lao Pilot

One of my routine rescues was of a Thai pilot who simply landed a little short of the pad he was going into. Flying H-34 side number 8770, he hit hard enough to cause the helicopter to roll backwards off the pad. It rolled over, rotor blades striking trees and bushes, ignited, and burned. When I arrived, all that was left of the helicopter was the usual steaming black lumps of charred engine and transmission. The tail section with the tail rotor blades had fallen away and remained intact. The pilot and crewman escaped unharmed. As the pilot boarded my helicopter for his ride home, he had in his hand the sheared-off FM antenna, its end splayed like a flower. He held it like a bouquet he might be taking home to his best girl. He was laughing, probably to relieve his fright and embarrassment, and at the absurdity of his "flower."

6 March 1972
H-53
FM Nakamoto

Proficiency check with senior instructor Captain James McEntee. More full-down autorotations and emergency procedure checks.

41

Worry Five

Drinking Wine on Duty with French Foreign Legion Para-jumpers.

12 March 1972
H-89
FM Lorenzo

After my experience of violating the drinking limit rule with Harry, I was always most careful to honor the Air America 12-hour bottle-to-throttle directive. I liked my job and did not want to be fired for something as simple as having a late drink in public. I made an inadvertent exception to that rule in Savannakhet. I had been assigned to carry a group of French plantation owners. I believe these fellows were former French Foreign Legion paratroops, but they still liked to jump. I didn't ask. I just carried them a few thousand feet up and looked away as they jumped out of my perfectly good helicopter that was not on fire or anything. After picking them up, I returned them to the terminal at the Savannakhet airport. I had never been there and never went there ever again. Air America helicopters did not usually work out of airport terminals.

While shutting down at the terminal, I experienced an unusual event. The hydro-mechanical clutch on my H-34 had seized up. The purpose of the clutch was to allow my rotor system to instantly and automatically release the engine from the rotor system should the engine fail, which would allow me to safely autorotate to the ground. If the engine did not release from the drive shaft there would never be enough back force from the rotor blades to turn the rotors and the engine, too.

Several times when I tried to back the throttle off to release the clutch, the engine stayed attached to the drive shaft. After conferring with Lorenzo, we decided that I would have to shut the engine off and let the whole system stop abruptly. I did not want to fly the machine even the half-mile to the Air America parking area. If the engine should fail during that short flight, the H-34 would have the glide ratio of a cement-filled safe. We would both surely die.

When I cut the fuel mixture, the engine lugged about three chugs, and the entire system came to a dead stop. The clutch had indeed frozen up. We were grounded at the terminal. Lorenzo said he would call operations for a new clutch. It was early afternoon, and I did not see how it would arrive before dark; I knew it would take Lorenzo hours more to change it out. I felt we were done for the day, so I decided to explore the airport.

I found a restaurant in the terminal. The Frenchmen who had just jumped out of my perfectly good helicopter were sitting in the bar drinking wine. They invited me to join them. I figured, *Why not? My helicopter is out of service until sometime tomorrow.* I sat with them and tried to speak a little French and pretended to follow the conversation. Over two hours, I enjoyed two glasses of wine with the Frenchmen. Then Lorenzo came into the lounge and said, "Captain, the helicopter is repaired."

I was dumfounded. How had he done that so quickly? It seems that as soon as he called in the problem, the company immediately launched a Volpar out to us with a spare clutch. Replacing the clutch was much easier than I had anticipated, and the job was done. As I stated before, these guys were good.

Now I had a huge dilemma to deal with. I had just flagrantly violated the 12-hour rule. Inadvertently for sure, because I did not expect the helicopter to be repaired that day, but now I had to fly it over to the Air America ramp area. But should I? I had been drinking. I'd had only a couple glasses of wine, but alcohol, nevertheless. I felt if I called in my predicament and confessed, I would be in big trouble, and probably get fired anyway. The specter of getting fired was always foremost in my brain, more than crashing or dying. I really liked this job.

After much anguish, I decided it was best to take the chance and fly the aircraft home. I swore Lorenzo to secrecy. For a while I feared the Frenchmen might brag to the chief pilot or operations about what a swell guy I was for drinking with them. But I realized that the French would never consider having a couple glasses of wine to be out of the ordinary, so there would be nothing for them to shout about.

Somehow, this particular incident never crossed my mind as a possible cause for me to be in trouble, as I flung my rotary wings towards Udorn for the fateful "Strela" meeting only two months later.

16-17 March 1972
H-63
FM Mondelo
Copilot for Joe Lopes

My logbook shows: "SAR Hobo 4.0 and SAR H67, Crane chase." No other details in my logbook or in my brain. "Hobos" were USAF A-1 Sky Raiders, usually sent out to assist in rescues of downed airmen.

19-20 March 1972
H-59
FM Campit

M.L. Morris my copilot. We fly four specials in the Louang Phrabang area.

MEANTIME, BACK IN THE REAL WORLD:

23 MARCH 1972

UNITED STATES STAGES A BOYCOTT OF PARIS PEACE TALKS. PRESIDENT NIXON ACCUSES HANOI OF REFUSING TO "NEGOTIATE SERIOUSLY."

At one LZ near Louang Phrabang, a Lao soldier presented me with the pith helmet of a North Vietnamese Army lieutenant. It seems the young enemy officer no longer had need of it. I still have it on my trophy wall.

NVA officer's pith helmet. PICTURE BY AUTHOR.

25 March
H-80
Koeppe my copilot

We fly more Crane Chase around Long Tieng and LS-272.

42

Dual Porter Crash At LS-69A

8 April 1972
H-86
FM Nakamoto

My crewman and I were sitting around LS-69A (*Ban Xieng Lom, D-8 on map*) waiting for an assignment. Word flashed to us that a Pilatus Porter (a short field take-off and landing airplane) had crashed nearby. I leapt into my helicopter and flew to the crash site, which was visible even before I lifted off, perhaps a mile away. We landed beside the crumpled Porter. Someone had pulled the pilot out and laid him on the ground. Dead. Nearby sat another Porter, intact. I thought the second Porter had landed in the rice paddy to help his buddy, but I soon found out that the reverse was true. The intact aircraft had been the first to land because of an engine failure. The pilot had no problem making a smooth, safe landing in the large, dry paddy.

The second Porter pilot flew over the first to make sure his buddy was Okay. He then pulled his aircraft hard up into a zoom climb to show off to his grounded buddy. He pulled up too fast, too hard, and pulled his Porter into hammerhead stall. His small aircraft plummeted from a couple hundred feet, slamming hard into the ground, killing him. My guess is that he died from sudden heart stoppage from the hard impact, because he was not badly messed up. I put my ear to his chest to make sure his heart had stopped. We loaded his body into my H-34 and hauled him the mile to LS-69A so his body could be returned to Vientiane.

Too sad.

Air America Pilatus Porter

From April 15th until May 20th,
The Number one song was
"The FirstTime Ever I Saw Your Face"
by Roberta Flack
It became number one for the year.

15 April 1972

Hanoi and Haiphong harbors are bombed by the United States.

24 April 1972

Captain Lloyd Randall is killed in his Porter when he hits a mountain, flying in bad weather from Long Tieng to LS-113.

29 April 1972
H-70
FM Ramos, R

We have an engine chip light. We shut down and investigate. Engine OK.

43

Worry Six

"I've flown in both pilot seats. Can someone tell me why the other one is always occupied by an idiot?"

–Anonymous

3 May 1972
H-80
FM Delacruz
Copilot for one of the "Fantastic Five."

A few days prior, I had flown with this particular pilot out of Louang Phrabang. He did something that day which absolutely appalled me, which set the stage for this day. At Louang Phabang the customer dispatched us on an urgent search and rescue mission. In his haste to get off the ground, this pilot did not do a magneto (mag) check. A mag check is something that should always be done after every start and should be done while sitting still on the ground. I expressed my concern. I hoped he would stop and do a mag check.

No. As we rolled down the runway at about 25 knots with our tail wheel off the ground, he reached up and ran the mag switch through its three positions, (from BOTH ON, to 1st OFF, back to BOTH, then 2nd OFF). I was aghast! Had we had a bad mag at that point, the engine would have quit abruptly. We would have made a violent right turn and rolled over into a fireball. This was not the first time he had demonstrated poor piloting skills to me. I did not like flying copilot for this guy.

This day we were carrying external loads out of a depression in the valley floor at Long Tieng. The pilot kept taking loads that were too heavy and doing hover-jump take offs out of the hole. There was no room for error. There was no place for us to go should we lose RPMs or have an engine failure. I felt very uncomfortable with what he was doing, and I asked him to carry lighter loads. He ignored my request. He did it several more times. I asked him again not to push the safety envelope so close to the edge. He again ignored me. Finally, I said, "Land this sumbitch and let me out on the ramp." He refused to do so.

I then took the controls and tried to overpower him, expressing physically my desire to land. With that, he finally realized I was serious, and let me out on the ramp. It was only about 30 minutes before the end of the workday, and perhaps I should have simply waited him out. I knew if I did that, I would be flying with him again the next day, and he would continue to put my life at risk. Instead of flying the rest of the day solo, or simply bagging it for the day, this pilot called operations and reported what had happened. We were both recalled to Udorn that night and told to report to the chief pilot the next morning.

Arriving at the chief pilot's office, I explained my concern. The senior pilot told his side of the story. As I recall, one of us got a letter of reprimand and the other a letter of censure. Whichever was the less serious, I was supposed to receive, but it never appeared in my mailbox. I learned later that the chief pilot knew some of the pilots were really bad, but there was nothing he could do about it.

The chief pilot did say to me, "If you don't like it here, then quit." It was not that I did not like it there; I loved this job. What I did not like was flying with dunderheads.

We Vietnam vets realized that some of these senior Air America pilots did not have 6,000 or 8,000 or 12,000 hours of flight experience as stated in their logbooks. They had 1,000 hours of flight experience six or eight or twelve times. They had stopped learning from experience. They made us Vietnam veterans very nervous. Before too long, the "Fantastic Five" expanded to be the "Nonsense Nine." Eventually the list reached eleven members, "Heaven's Eleven." Writing this I realize there was one more … a full dozen dunderheads. Amazingly enough, every one of these piss-poor pilots survived the war. I am sad to say that one of our Vietnam Marines made the list.

I will never divulge this list. Do not ask.

8 MAY 1972

SOUTH VIETNAMESE PILOTS ACCIDENTALLY DROP NAPALM ON A VIETNAMESE VILLAGE. THE PHOTO OF THE BURNED, NUDE YOUNG GIRL FLEEING IN FEAR BECOMES YET ANOTHER ICONIC IMAGE OF THE UNPOPULAR WAR.

44

Another STO
The Taj Mahal and Nepal with Sweet Syn

I travelled to Bangkok and joined up with my favorite Pan Am stewie, Sweet Syn. We jetted over to India to visit the Taj Mahal. There is nothing more romantic than visiting the Taj Mahal by the light of a full moon. I highly recommend it.

Picture of Syn and me at the Taj Mahal.
We visited that night under the light of a full moon.
PICTURE FROM AUTHOR'S PHOTO ALBUM.

We also visited the Red Fort across the river. While walking around the Red Fort, we found ourselves walking beside an Indian couple. When the ever-curious Syn asked the man questions, he ignored her. When I repeated the same question, he answered me in detail. This happened several times. The Indian man simply could not acknowledge that a woman was speaking to him.

Later this same trip, we also flew into Katmandu, Nepal. Unfortunately, there was a huge thunderstorm with huge hail just as we arrived. The storm disrupted all the airline schedules for the next few days. We spent most of our remaining time in Katmandu working to arrange our departure, as we both had flight schedules to meet. Consequently, we saw little of Nepal's capitol city and nothing of the countryside.

I bought a copper replica of a Buddhist temple horn. I call it my Yeti protection horn. I blow it once in a while, usually on New Year's Eve, to keep the Yetis away. It seems to work, as I have yet to see a Yeti in my neighborhood.

Picture of Tibetan temple horns
from Reader's Digest magazine, October 1972.

Syn eventually married a Xerox executive and replicated herself.

I heard a story about one of our senior pilots visiting Katmandu. I believe it was Charlie Weitz. It seems he went there alone and stayed at the Hilton. One morning, he went down to the dining room for breakfast. There was only one other person in the dining room, and that distinguished looking person invited the pilot to join him. They enjoyed a pleasant meal together. The Nepalese man was charming, educated and erudite. He asked the pilot all about himself and what he did for a living and so on. Finally, the pilot thought to ask his host what he did.

"Oh," said the Nepalese man, "I'm the king."

45

Swamped!

Flight Details not recorded.

Another day I was flying out east of Pakxe on the Bolivens Plateau (*N-19 on map*) supplying an outpost of local troops. It was late in the day and there was a rumor that the North Vietnamese Army troops were planning to attack a nearby compound. No one knew which one. The troops must have heard that rumor, for when I departed the LZ, several of them requested that I take them out with me. I denied them. A few minutes later as dusk neared, I made my last delivery to that LZ. It was getting dark, time to return to Pakxe and secure for the day. The troops, in a panic, would not take no for an answer. They swarmed aboard my helicopter. My FM told me we were grossly overloaded. I told him to tell some of the troops to get out. He said, "I have, captain, but they will not." I evaluated the option of shooting them, but thought the better of it. They might shoot back; we were grievously outgunned.

I pulled the helicopter into a high hover. It could not hold a hover and slowly settled back to the ground. This was scary. If I could not get them out, I could not get myself out, either. I decided I must to do the hover-hop technique that I had practiced a few times in the recent past, albeit with no passengers on board. I could see a clear path to escape, even though I knew the hop over tall trees would cause my RPM to disintegrate towards lethal minimum turns. I pulled in maximum allowable RPMs and then added few more for good measure, over speeding the engine and the rotor system. I yanked the helicopter off the ground in a large hop over the tops of the trees. As the RPM disintegrated, I pushed the stick forward and slid the machine into forward flight, gained translational lift, and flew away to safety.

My FM told me we got out with 33 local troops on board. Normally eight would have been a full load. So far as I know, the NVA did not attack any outpost that night.

46

Henri's Roadside Bar and Grill, Pakxe, Laos

After a hard day's work around Pakxe, putting our lives on the line in "the battle against the red tide of communist aggression," we often felt we deserved a little reward. In the village of Pakxe, we gathered nightly at Henri's Bar and Grille for drinks. Pakxe was the regional big city and as I recall the population was numbered only in the hundreds of people. The entire population of Pakxe province was only about 60,000 people.

Henri's was a simple affair, sitting between the highway and the Mekong River. The whole business was a simple square box, open to the busy street. Two plain wooden tables, each less than three feet square, accompanied by four equally plain wooden chairs, served as the totality of the sparse furnishings. These furnishings had no paint or discernable finish of any kind, nor any hint at ever having had a finish of any sort. They sat about two feet lower and about five feet from the edge of the highway.

There was a constant bustle of traffic, dust, and noise as pedestrians, bicycles, ten-wheeler cargo trucks and tuk-tuk taxis passed by. Behind the noise and bustle there was a certain romantic ambience of being in this quiet, exotic, riverside Laotian village so far from home. Unlike Vietnam, here we had no fear of café bombers, sappers, or snipers.

Henri's food and drink offerings equaled the furniture for variety. There was no grill and no food. For drinks, Henri offered only Mekong whiskey, the local Laotian beer, and French wine – Beaujolais

– in a can. Two bucks – a thousand kip – would buy a can of wine the size of a 12 oz. soda pop can. It was drinkable and better than the other choices. We would sit around, swap lies and talk about the events of the day.

One evening several of us were sitting there beside the highway when someone recognized an old buddy walking by. “Hey Jim!” someone shouted. Jim joined us for a can of wine. Several of us knew Jim from Navy flight school or having flown with him in Vietnam. It was great fun to hail an old friend in such a strange atmosphere. He was one of us. Or, he had been. Jim was no longer a helicopter pilot. He was now the customer “Greek.” We loved him anyway.

A few weeks later, the situation repeated itself. I don’t remember the next fellow’s name, but another of our Vietnam H-34 pilot contemporaries turned-customer walked by Henri’s as we sat there drinking. He also came over and joined us for a drink, then went on his way never to be seen again. We joked that if we sat at Henri’s Café long enough, everybody we knew would eventually walk by.

In 2015 I read this in the book *Charlie Wilson’s War*, about the CIA officers involved in the build up in Afghanistan to counter the Russian invasion of that country: “Some [had been] pilots for Air America and they were crazy, but that’s what you want on your side.”

A Round of BJs

One evening five of us were sitting around Henri’s crude tables drinking our cans of wine. I am sorry to say that I do not remember who any of the other guys were, but I know Gary was not there. A beautiful young Lao lady joined us. We all knew her and called her “Number Nine,” also known as the girlfriend of one of the married Air America fixed wing drivers. Rumor was she had even had a baby with him, but that did not stop her from plying her trade. A girl has to make a living, after all, and the pilot was not supporting her so far as any of us knew.

After a short while, Number Nine and one of the other pilots had a hushed side conversation. They got up and walked catty corner across the street to her shabby four-story hotel, hand in hand. After a short absence, they returned; the pilot had a smile on his face. They rejoined the table. Soon Number Nine took off with another of the guys to her hotel room. He, too, soon returned to the table with a smile on his face. Pilot number three soon departed the table with Number Nine, and got his business taken care of, too. Then pilot number four. It was my turn. I walked across the busy intersection with Number Nine, up to her room – number 9 (hence her nickname) and had my needs met, too. When I returned to the table, everyone else was gone. I caught a taxi back to the company hostel.

Another BJ

A few weeks later, I happened to be sitting alone in the company hostel living room, which I rarely did. Number Nine entered looking for her boyfriend, she said. I informed her that he was not in Pakxe at the time. She sat next to me on the couch. After we chatted for a while she put both her hands on my thigh. There was no mistaking that invitation. I enjoyed another BJ right there in Mr. Lee's hostel living room.

When all was done, I took out my wallet. I gave her everything in my wallet, explaining that was all I had with me at the time, which was true. She happily accepted my 200 kip, the equivalent of 40 cents U.S.

47

A Disastrous Special Mission

Tony Byrne Shot Down
Freddy Alor Dies

19 May, 1972
H-92
FM Adger
I fly copilot for Joe Lopes

This story explains why I note every special mission I ever flew. This one turned to worms.

I was flying copilot for Joe Lopes, on an insert north of Pakxe. Joe and I were Search and Rescue for Tony Byrne in the lead H-34. Carpenter was copilot for Byrne; Freddy Alor was their FM. The mission was simple: to insert a team of local troops on the edge of a plateau near Khong Sedone (*M-18 on map*). At the pre-mission briefing, we were told that the LZ in the area was reported to be quiet with no reported enemy activity.

Tony approached the large LZ. It was on a peninsula of the plateau, a large field ringed on three sides by tall trees, with cliffs just beyond the trees. All looked quiet. Tony circled down and made a normal approach. As he flared out at the bottom to complete the landing, he took a heavy burst of AK-47 automatic weapons fire right into the nose of his machine, right into the 1,500-pound bullet-stopping engine. Tony instantly cobbed maximum power to his engine and leapt his H-34 off the ground. He jumped his aircraft over the tall trees, heading east. Once he cleared the trees, he was screened from further fire.

Just as he cleared the trees his engine quit.

There was no turning back to the LZ. It was crawling with enemy soldiers. Tony quickly commenced an autorotation and expertly glided down towards rice paddies 700 feet below. Fortunately, there were hundreds of acres of dry, empty rice paddies beneath him. He made a smooth landing.

On touchdown, the right main gear of Tony's helicopter collapsed, probably riddled and weakened by bullets. The helicopter rolled to the right. The tilt was not enough for the helicopter to roll over, but it was just enough for the blade tips to sweep the ground. Joe and I both sighed big breaths of relief. It looked like we had escaped without a serious problem.

FM Freddie Alor, one of the very best, riding in the crewman's seat, had always expressed a fear of dying in a burning helicopter. His fear got the best of him. He panicked and tried to run away from the almost-crashing helicopter. We watched with despair as he ran directly into the decelerating blades. One of the blades caught him on the top of his head, striking him down violently. Freddy's fear cost him his life. I remember Joe Lopes saying with compassion "Oh, Freddy!"

One of the Lao troopers in the passenger compartment, following Freddie's lead, also panicked, and ran out into the blades. He caught the very last bit of motion from a slowing blade right across his back. It whacked him violently to the ground, too. Joe and I landed beside the disabled helicopter. I unstrapped, and jumped out to help the troops and other crew into our helicopter. Carpenter jumped down from his seat beside Tony to help me. We loaded the two bodies. Freddie was beyond help. The blade had caught him right at the hairline, removing his cranium. His brain was exposed, smushed. I'm sure he died instantly.

Because of the urgency of the matter, we had no time to be gentle with the Lao trooper. We were sure that we were going to receive heavy automatic weapons fire and mortar fire from the plateau above at any instant. Carpenter and I each grabbed an end of the trooper, and tossed him into the helicopter like a sack of potatoes. I'm sure that he was already doomed, but our throwing him aboard the helicopter did not help his condition. Carpenter and Tony jumped aboard and I climbed back into my copilot's seats. We returned to Pakxe.

All that was left for us to do was write our reports of the incident. We heard later that the Lao trooper died of his injuries. The next day we flew over the site, and all that was left of the helicopter was the usual: charred engine and transmission, covered in black ash, the rotor tips and the tail pylon still intact. The enemy had torched it during the night.

This was a common sight to us. Scratch yet another H-34.

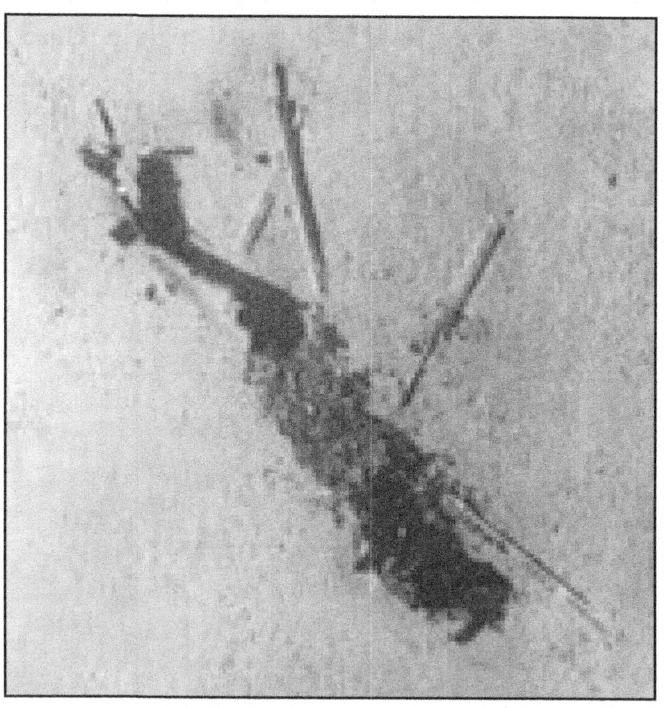

A burnt up H-34 on a beach in South Vietnam.
PICTURE COURTESY OF WWW.POPASMOKE.ORG

Dr. Joe Leeker wrote about this in his *Aircraft of Air America*: (See bibliography).

"Fate: aircraft received heavy small arms fire during approach to landing at coordinates WC8021, near Khong Sedone (LS-289), Laos on 19 May 72; the engine failed and the aircraft made autorotative landing in a tree-congested rice paddy; F/M Alfredo J. Alor was killed, when he was struck by the main rotor blade; one passenger was injured; the aircraft received substantial damage to the right main landing gear; as the site was not secure, it was hit by mortar fire while the crew was being rescued."

[Author's note: There were no trees. We did not take mortar fire. The enemy destroyed the machine that night because it sat unguarded overnight.]

48

More Cheap Thrills

I Lift a Heavy TACAN Box at LS-272

22 May 1972
H-45
FM Adger
Copilot Lee Emery

The Company was building a fallback base at LS-272 should Long Tieng be taken by the NVA. Technicians were installing a TACAN navigation radio atop a steep, jungle-covered hill near LS-272. A large metal shipping container full of vital electronics sat on the ground at the base of the hill. It needed a lift to the top of the hill. A U.S. Army Sky Crane (CH-54) had been doing this heavy lift work, but something had gone amiss the afternoon before, and the sky crane had crashed on top of the hill, killing the crew of three. The Company needed to get the job done, and I was asked to do it. I asked the weight of box. No one knew. I assumed it was heavier than I could lift.

I at first told the customer no, but after I thought about it for a while, I realized that I might be able to do it. I coordinated with the customer. I told him I needed to work around the area for a while until my fuel had burned down to minimum. If we then unburdened the ship to its lightest, I might possibly be able to lift the box.

I flew around and made several more deliveries to local villages. I flew up and surveyed the hill where the box needed to go, to get an idea of what I was getting into. When I had burned down to minimum fuel, I returned and landed near the big box. I had FM Adger unload everything that was not attached to the helicopter. I told him to stay on the ground. I told copilot Emery to get out and wait for me. He seemed only too happy to cooperate.

I hovered over the big box. A crewman hooked it onto my external load hook, and I coaxed my engine to lift it. It came off the ground easier than I expected. I flew a large 360-degree right spiraling climb so that as soon as I had climbed high enough to see the place where the box was going, I was on final approach for that spot. I eased that big box right down onto the cement pad that had been prepared for it. It was a little disconcerting to see the remains of the sky crane scattered all over the hilltop where it had crashed the day before.

I knew that I had taken a big risk doing this lift, but I did it anyway. I liked the challenge and needed the adrenaline rush that I got when I did something a bit "outside the box." I returned to the base of the hill, retrieved my crew chief, Emery and loose equipment, and returned to LS-272 for fuel. We continued to fly routine missions for the rest of the day.

49

The Press

Early in the summer of 1972, well-known war photojournalist Anne Darling showed up in Udorn, hung around some the bars we frequented, and pumped the fellows for information about what was going on upcountry. We all had utmost disdain for these snooping, newsy kinds of people from our experiences in Vietnam. I wrote about this in my first book:

The Adventures of a Helicopter pilot,
Flying the H-34 in Vietnam for the United States Marine Corps

"While we flew all around I Corps, we often carried reporters and photographers. We disliked carrying these people because for each press person we carried, we felt it was one less Marine we could carry, or that much precious cargo needed by our Marines that we could not carry. This was especially important to us when we were carrying wounded Marines out of the field. It got to the point that whenever we saw someone pointing a camera at us, we flipped off the photographer. Somewhere, buried deep in the Marine Corps archives, there are hundreds, perhaps thousands, of pictures of Marine Corps helicopter pilots flipping off the photographers of those pictures."

The attitude carried over to Laos. We often felt that much of what the press reported was not factual. There were times with Air

Am, too, that we knew what was really going on was nothing like what the press reported. Sometimes the reporters lurked in bars and picked the brains of pilots returning from missions. Sometimes they reported as fact what they overheard in the bars. One time I refused a ride to a reporter who wanted access to an area where a battle was happening. The next day we read his first-hand article about the battle. I believe he never actually got there.

So of course, the fellows did the natural thing. They pumped Ms. Darling full of beans. Ms. Darling wrote a lengthy article for October 1972 premiere issue of OUI magazine entitled, "CIA Super Pilots Spill the Beans!" I know all the guys she interviewed and can attribute each quote in the article to a certain one of my friends. Nevertheless, the article is interesting, somewhat true, and gave me the title for this book.

I was not around for this caper, but wish I had been.

The entire article is recreated in Appendix A. It has been formatted to fit this book. No words have been changed as per my agreement with the current owner of **oui** magazine, who most generously gave her permission for me to use this article and the cover of the magazine in its entirety.

50

Worry Seven

I Insult Two Senior Pilots

And then there was the little incident in the bar at Tango, but no, that couldn't be a reason for firing me, could it? But could it be a final deciding factor ... Arrgh! My mental cooling radiator was well into the red arc. It seemed I could feel steam coming out of my ears!

Shortly before a recent trip upcountry, I was in the company bar, drinking with a couple of the older pilots, old timers who had been around forever, fixtures in the company. When I had entered the bar, one of them offered to buy me a drink, an offer I found hard to refuse. I asked for my favorite, a JWB scotch over ice. They seemed to be celebrating something. I drank the drink, and then to be sociable, I asked, "What's the big occasion?"

"We just got fired for breaking the 12-hour rule," one said.

At that point, because of my not-too-recent transgressions of the twelve-hour rule, I felt very guilty. It was enough that I had blatantly broken that same rule myself, but here and now I felt I insulted two senior pilots who were just fired for the same, yet a much more benign, offense. They had simply been seen having a glass of wine in the club with their dinner. My offense was much more serious. Did the senior pilots know about my transgressions? Had they, in a fit of retaliation, reported me to management?

Was I next to be fired?

Worry Eight
All of the Above?

Maybe it wasn't any one of these things that was leading to my firing. Perhaps it was the accumulation of all these things. ARGH! My mental anxiety gauge and my brain over-temp gauge were both banging against the stops of their red arcs. I could feel my brain boiling and pressure squeezing my brain.

Maybe I was unfit to be an Air America pilot. I would have to return to the states a failure. I would have to become an alcoholic to drown my disgrace and relieve my sorrow. I would die young, broke, broken, and all alone.

51

Sleeping in the Cockpit While Flying

One day after lunch in Vientiane, I was flying north towards the area of "The Rockpile" (not the same Rockpile as in Vietnam). I began to get really sleepy. I fought it as hard as I could, opening my window, shifting my position around as much as I could. I even slapped my own face and pinched myself, all to no avail. I fell asleep anyway. I leaned forward in my shoulder straps, and my head fell forward as I slipped into sleep.

As soon as my chin met my chest, I heard the engine quit! The ear-splitting silence awakened me in a panic. I quickly placed my hands on the controls to shoot an autorotation and simultaneously looked to the ground below for a place to land. Before I could even lower the collective, my eyes and ears told me the RPM was within normal range. A quick scan showed me that all the other instruments were within normal limits, too, and none of the 13 warning lights had illuminated. The engine had not quit.

What was that noise?

I realized when I passed from awake to asleep, my ears dropped off line, meaning what I heard while I was awake became much quieter when I slept. This is what caused the noise of the engine to greatly diminish.

With a little practice, I was able to drift down past the engine-failure, ear-splitting silence and get into sleep mode. I felt it was better to sleep for ten or fifteen minutes than fight it for an hour or more and be groggy and lethargic. After a short nap, I was awake, alert, refreshed and ready for a challenging afternoon. It got so that on long legs I routinely took these little naps.

I developed a regular technique of tightening my shoulder straps, getting the helicopter neatly trimmed, and then simply relaxing and letting George, the autopilot, drive the machine. Because George could only do the heading and the pitch, (the altitude mode of the device was inoperative), I had to be sure that I had everything quite stable. With the H-34, that was fairly easy to do. I also did this napping only in cases where I had a long leg in front of me with no mountains or obstacles to avoid. I also knew that there was little air traffic in my area, and I usually knew who and where that traffic was. I also knew that the sleep I was enjoying was light. If something should malfunction in the helicopter, I would be instantly ready to handle it. At least that is how I felt about it. I never really had the chance to prove it, as I never had a mechanical problem with any H-34 while doing this.

There was the story of an Air America H-34 pilot who took quite a nap and awoke to see a huge lake in front of him. "What lake is that?" he wondered. He checked his map and memory, and could find no lake of any size in the direction he was supposed to be going. It was then that he looked at his compass and realized that he had been flying a course of 130 degrees, not his intended 310 degrees. He was 180 degrees off course. In his sleep, he had completely overflown the complete width of North Vietnam. The lake he saw in front of him was the South China Sea. He scurried out to sea, flew down the coast and made it to a U.S. military base with his low-fuel warning light on. There he took on fuel and flew back over the mountains into Laos without any repercussions.

This is a confirmed legend in Mule's books.

52

Special Flight into Cambodia

An Engine Chip Light

29 May 1972
H-70
FM Ramos
Copilot Ellis "Lee" Emery

I was flying out of Pakxe when the customer laid out a mission that required four H-34s to go south into Cambodia. We were to insert a recon team of local troops a short distance into that country. My job, with my empty ship, was to be Search and Rescue for the other three ships. The brief assured us it would be a routine mission. (The customer always said the mission would be routine.) We flew our flight of four H-34s south across the border and landed at the disused airport near Phumi Narung (*M-22 on map*).

I don't remember why we landed at that isolated airstrip, but all was still. Still was always spooky, but in this case still was safe. After another short brief on the ground, we departed, heading southwest. A few miles out, my engine chip detector light illuminated. I told the flight leader that I was returning to the disused airport to check out the engine. Just for practice, I rolled the engine back and shot a full-down autorotation to the runway. We had to wait a few minutes for the engine to cool enough for FM Ramos to pull the sump plug. When he did so, he found no metal bits on the chip plug, which meant that the engine was OK. By the time we got the plug pulled, checked, and back into the engine, the other helicopters passed overhead, returning to Pakxe. We joined up on them and flew home. It had indeed been a very routine insert.

This turned out to be my last H-34 flight for Air America, my last flight as a captain, and my last flight with a radial engine at Air America.

Those of us who flew in aircraft with radial engines love the sound of those big engines starting up.

ROUND ENGINES
(Author unknown)

DEDICATED TO ALL THOSE WHO FLEW BEHIND (or above) ROUND ENGINES

"We gotta get rid of those turbines; they're ruining aviation and our hearing...

A turbine is too simple-minded; it has no mystery. The air travels through it in a straight line and doesn't pick up any of the pungent fragrance of engine oil or pilot sweat.

Anybody can start a turbine. You just need to move a switch from "OFF" to "START" and then remember to move it back to "ON" after a while. My PC is harder to start.

Cranking a round engine requires skill, finesse and style. You have to seduce it into starting. It's like waking up a horny mistress. On some planes, the pilots aren't even allowed to do it.

Turbines start by whining for a while, then give a lady-like "poof" and start whining a little louder.

Round engines give a satisfying rattle-rattle, click-click, BANG, more rattles, another BANG, a big macho FART or two, more clicks, a lot more smoke and finally a serious low-pitched roar. We like that. It's a GUY thing.

When you start a round engine, your mind is engaged and you can concentrate on the flight ahead. Starting a turbine is like flicking on a ceiling fan; useful, but hardly exciting.

When you have started his round engine successfully your plane captain looks up at you like he'd let you kiss his girl, too!

Turbines don't break or catch fire often enough, which leads to aircrew boredom, complacency and inattention. A round engine at speed looks and sounds like it's going to blow any minute. This helps concentrate the mind!

Turbines don't have enough control levers or gauges to keep a pilot's attention. There's nothing to fiddle with during long flights.

Turbines smell like a Boy Scout camp full of Coleman lanterns. Round engines smell like God intended machines to smell."

53

A Newer, Much Bigger Worry

As it turned out, all my worrying was for naught. As I cruised for home, I heard many other pilots call in with their position reports. It soon became obvious to me that everybody upcountry was being called back home at the same time, like a flock of homing pigeons all headed for the roost. Relief! At least I now felt comfortable that the reason for the recall was not personal. What was up? Now there was another mystery to puzzle over. Why this unprecedented general recall of all aircraft?

About a half hour out of Vientiane, I made a position report. Another pilot made a position report immediately after mine. He reported being in exactly the same position I was in. My head swiveled around quickly in a frantic visual search but I could not see another helicopter. I radioed the pilot, Jesse Hagerman, and I asked him his exact location so I could avoid him. It has been well established in aviation that a mid-air collision will ruin your entire day. Jesse replied, "Look on your right wing." There he was, flying tight formation on my right rear. He had sneaked up close behind without telling me. I was not pleased.

Upon returning to Udorn, all pilots received instructions to report to the company theater at 7 p.m. for a general briefing. Rumors were flying; tension was up. Nobody knew what to expect, but we all were most curious. Before the appointed time we all gathered at the theater. We saw a technician with an electronic sweeping device checking the theater for "bugs." This was mysterious and unprecedented. This really was going to be interesting, and probably most secret.

Yes it was.
Both.

At the briefing, we learned that the enemy in our area was reported to now have Strelas, the small, portable, shoulder-fired heat-seeking missiles. We knew the enemy in Vietnam had been using these missiles for some time, and we always knew that it would be just a matter of time until the enemy in our area would have them. The enemy in our area now had the dreaded Strelas.

Soviet soldier aims Strela missile, (FROM WIKIPEDIA)

At this time, I had a flashback to Vietnam – six years prior. I was flying north of Phu Bai (*Q-16 on map*) for the Marines Corps in the fall of 1966. I knew that our jet fighters had small heat-seeking missiles called Sidewinders that could shoot down enemy aircraft by tracking in on their thermal signatures. I had taken one look at the situation in Vietnam and said to myself, *If the enemy had only a few portable heat-seeking missiles, this war would be over in two weeks.* At the time, I never shared this thought with anyone for fear of it being prophetic.

At the secret briefing, we were told to be on alert for the missiles. Sure we would. But how do you watch constantly for something that can reach Mach 5 in two seconds, something that has the capability to go from the ground and bite your navel from the inside faster than you can say, "Oh, shit !" and fly your aircraft, too?

It was several months before anyone saw one of the missiles fired. During this time my group of close friends had several long conversations over late night drinks about risk, death, evasive maneuvers, et cetera, concerning the missile threat. We had been big fans of the Israeli helicopter pilots when they had learned how to protect themselves from Arab missiles during the 100 Hour War in mid-1967. The Israelis had learned to turn into an oncoming missile and to dive for the ground. The streaking missile could not turn fast enough to follow them, and the radar control of the missile could not track a helicopter against the background clutter once it was lower and slower than the missile. So far as we knew, no Israeli helicopters were ever shot down by the Arab missiles. But this procedure was predicated on seeing the missile headed your way. You had to see it launch.

About the only thing we all agreed on was that we were all going to stick around until the first one of these things was fired at one of us, hoping that we ourselves were not the target of that first missile, and then we were going to quit. Discretion being the better part of valor and all that jazz.

It was during this time that we began to better define the "fear/greed ratio." All the time we were flying, we knew that this kind of flying had certain risks ... small arms fire … 50 caliber fire … 37mm fire … mortars in zones … rocket-propelled grenades … rockets … mines … Migs … mid-airs … weather-related hazards …claymore mines … engine failures or other mechanical failures of our machines … death by stupidity – our own or someone else's – and a thousand other factors that made life interesting. But this new missile threat threw a big unknown into the fear/greed equation. Up till then, we felt that we were making enough money to mollify the fear. All my close friends had been through Vietnam, where flying was a whole lot hairier than flying in Laos for Air America.

The pay and bennies at Air America were much better. I figured that the risk of flying (pre-Strela) in Laos was one tenth of that in Vietnam, and the pay was nearly five times better. Do the math – this job was 50 times better than flying for the Marines in Vietnam. (All that and hot and cold running airline stewies, too!) So as long as neither the fear nor the greed got too out of hand, we felt safe.

Of course, greed could kill us too, but few of us ever got out of hand on the greed side. We got paid an extra $50 per landing, or per hour, whichever was more, for the specials. They were usually well-planned, thought-out missions, and usually went according to script; rarely did anyone get killed or even come home full of holes. There were incidents … see above.

My friends and I all agreed that we would stay with this operation until the first Strela was fired. Then we would retire. Gary and I figured we had made enough money by now that if we played it right and invested wisely, we would never have to work a regular job again.

> About the year 2000, I found the website to adjust for inflation, www.usinflationcalculator.com/. I punched in the amount of money I was earning in 1972, about $38,000 per year, and what it would translate to today (July, 2019): almost a quarter of a million dollars. That is more than 1,100 dollars per flying day. This was at a time when the dollar was strong, and before Nixon took the U.S. Dollar off the gold standard. I remember being in Tokyo in 1972 and getting 454 yen to the dollar. Today (3/18/2016) the exchange rate hovers around 100. No wonder we felt like we were doing well. And had money to spare.

Gary and I limited our personal expenses to $500 per month. This included rent, food, and our monthly STO travels. We were able to do all these things for less then $500. The rest we socked away into blue chip stocks.

54

Close Calls

"Did you ever have any close calls?" is almost always the first question I get whenever I tell people I flew helicopters for 32 years. My immediate response is, "Yes, of course." I read recently a report by an old-time WW2 fighter pilot. When asked that same question, his response was, "Every time I take one of these sumbitches off the ground it's a close call!" I can honestly say I came up with the same phrase before I read it in that WW2 pilot's story. And ... I can beat his story.

One day I had a close call before I even started my engine.

I got my aircraft assignment from operations at Udorn. I walked out to my H-34 on the ramp, did my pre-flight inspection, slipped into my NOMEX flight suit and climbed into the machine to prepare for a normal six-day trip up country. I began the pre-start check-off checklist:

First item: Battery switch ON I pushed it down.

Second item: Electrical driven fuel pump ON I toggled it up.

I checked the fuel pressure gauge: fuel pressure within normal limits.

What wasn't within normal limits was the spray of raw fuel jetting from behind the instrument panel. The gage was known as a "direct reading" gage, which means that a small fuel line from the fuel pump was connected directly to the back of the gauge. That line had broken off at the back of the gage. Raw fuel of the highest octane began to spray all over me and the cockpit. The smell of raw fuel filled my nostrils. It quickly filled the cockpit with a mist of highly volatile fuel-air mixture. Raw fuel in the proper fuel-air mixture is more explosive than dynamite!

Fuel was dripping down the back of the instrument panel, past several switches and onto the radio console besire my left knee. None of the radios was yet turned on, but the relays and busses to them might now be hot with electricity. The smallest spark would ignite that fuel vapor. I knew I would not survive the explosion. I also knew that even if it did not explode, as soon as the raw fuel reached any ignition source there would be an instant fireball.

I had to make an instantaneous decision. Quicker than you can snap your fingers, I had to choose between abandoning the helicopter or trying to correct the situation by switching the battery switch to the OFF position. I knew that if I jumped out of the helicopter, the fuel would continue to flow. If (when!) the helicopter caught on fire, the magnesium alloy would soon ignite and in 15 seconds the entire helicopter would be violently consumed. Jumping out would have also put my FM, at risk. He stood beside the helicopter with a fire extinguisher at the ready, as he always did, in case of an engine fire on start. What we were facing here was much more dangerous that an engine fire on start.

Did I want to leave my FM with an explosive situation while I ran away? Or did I want to risk being some of the ashes? Should I bail or stay? I might escape with nothing more than fuel-soaked clothing, but the helicopter might burn up, and we both might be badly hurt or killed by the exploding helicopter.

No, I could not bail out and put my FM at risk. I never had an FM I did not like and respect. I did not wish to put him at risk. I quickly snapped the battery switch up to the OFF position. There was no spark.

It seems I made the right decision.

On our evenings at home plate, whoever was there got together and we went out for dinner. There were several options. The U.S. Air Force base had a couple of restaurants, our Air America rendezvous club had good food, and there were plenty of restaurants in Udorn. One of these was a Chinese restaurant called Five Sisters. Whenever we went to Five Sisters, we always ordered the ribs for an appetizer. They were perfectly sweet, delicate and tasty.

After a year or two an article came out on the front page of the local English language newspaper that the restaurant was selling dog meat. Photos showed live dogs in cages and dog carcasses hanging in the kitchen. Five Sisters was shut down by the local authorities.

After about two weeks, money changed hands, fines and bribes were paid. The restaurant reopened. We returned to our favorite Chinese place. Ribs were still on the menu; we ordered them again. They were again the same small, perfectly tasty ones we had before. We knew full well that we were eating dog ribs.

55

Bell Hueys

During early 1972, as things began to heat up, we began to fly double captains on almost all missions because we were getting shot at more and more frequently in more and more locations. Then the Strela threat appeared. After a while, the Company realized that it was costing a lot of extra money to have two captains in every cockpit. Their answer was to reorganize the seniority list and bust a number of junior captains back to first officer. In aviation, like the military, seniority is everything.

Junior enough to be affected, I was busted back down to first officer. I was given a choice. I could stay first officer in H-34s or I could retrain (still as a first officer) in what we called "Bells," the civilian versions (204 and 205) of the Huey, built by Bell Helicopter Company. It was a hard decision. There were pros and cons to each avenue.

I did not want to subject myself to more hours of flying copilot with the "Nonsense Nine" (the by-now expanded version of the "Fantastic Five"). I thought it would be good for my career to get some more turbine time. Turbine engines were becoming dominant in helicopters because of their lighter weight and better reliability. I had a bit of turbine experience in a Bell Jet ranger during my short time on Alaska's North Slope prior to Air America, but I felt I would need more turbine experience if I wanted to continue flying after Air America.

I wrestled with the fact that the Bells put out a larger missile-attracting thermal signature, and they worked in the area most hotly contested in Laos. Somehow I convinced myself that flying Bells would be Okay. Another advantage to flying Bells was that the workdays were long, and pilots usually got all their flying time in each month in only 10 or 12 days. In the Bell division, STOs were ten days; lots more time off.

I was not too happy about a huge cut in pay, but I decided to ride out this change and continue flying with Air America. It never occurred to me to quit over this. I was part of a big machine, I loved the work, and I was sharing the experience with a group of great buddies. I was along for whatever fate brought ... at least for a while longer.

At this point, I submitted all my documentation and the Company arranged for me to obtain an official Lao pilot's license because some of the Bellswere registered in Laos and had Lao registration numbers that started with XW.

2

Signature du Titulaire (VII):

NOM (IV): COLLIER,
Prénoms (IV): WILLIAM FRANK
Né le
à U.S.A.
Nationalité (VI): AMERICAINE
Domicile (V): C/O AIR AMERICA INC.
VIENTIANE, LAOS

Délivrée à Vientiane, le (X) 22 NOV 1972
Signature du fonctionnaire
délivrant la licence (X):

Cachet du Service
délivrant la licence
(VIII et XI)

We never thought of ourselves as actual mercenaries, and we were doing our work partly out of patriotism to democracy and the American way. But flying a Lao-registered aircraft with a Lao pilot's license in Laos for money made us just about as mercenary as you can get. The only exception was that we were being paid by the U.S. government and not the Lao government. This would be a fine point to debate with North Vietnamese captors should I ever have the misfortune to be taken prisoner.

Rumors were that Air America was an offshoot of the Flying Tigers of WWII. To enhance their chances of survival, should they be shot down, the Flying Tigers carried "blood chits." These were U.S. flags sown onto the back of their jackets. In several local languages on the flag was a statement that whoever helped this pilot to safety would be rewarded in gold. These Flying Tiger pilots also carried small bars of gold that they could use to help buy their freedom.

This custom prevailed at Air America, but it became a bit perverted and was at our own expense. We bought ID bracelets made of gold links. The idea being that the links could be cut off and these smaller bits could be used to buy favors, should the need arise. It did not make sense to me because if you began to flash that much gold around the tribes-people, I thought they would simply rob you or even kill you for the bracelet. One of these bracelets represented the equivalent of a decade of earnings for them. I bought one anyway, the norm while I was there.

They were simple at first, just being a simple single chain holding a small plaque with the pilot's name engraved under a depiction of Air America captain's wings. I personally bought a double-chain bracelet that weighed 454 grams. It cost me about $454 in 1971. I had my Air America wings embossed on it with my name underneath. On the back I had my blood type etched. A few years later, I got into a financial jam and had to sell the bracelet. I sold it for about $2,000, a good return on my initial investment. I liked my bracelet. I wore it at all times, especially on STOs to Bangkok or Hong Kong. It was a trademark of our business and marked us as fearless mercenaries. The young ladies were impressed. Negatively at times, I am sure.

You can see my gold bracelet in the photo from the Hong Kong Post Herald describing the Up Club spaghetti-eating contest. *(See p. 235)*

Anytime you get a bunch of high-spirited, high-testosterone driven young men together, you get a competition. It culminated in one of the fellows buying a gold bracelet with a plaque about three by four inches that had three chains to go around his wrist. It must have weighted 1500 grams, about a pound and half. It was grotesque. He never wore it in public, just a few times in the bar to show it off.

I found that wearing the gold bracelet on my wrist threw off my dart game so I began removing it from my wrist while playing darts. After a while, my game again began to deteriorate. I reasoned that my arm had adjusted to the weight of the bracelet. I put it back on, and my game improved. I had adjusted to the weight.

Some fellow also bought "baht chains" gold chains in which each link weighed the same as a one-baht coin. They hung several small Buddha amulets on the chair to protect them from danger. I saw how well the Buddha protection worked for the Lao troops, but the amulets seemed to work for our fellows. I actually was given a Buddha by a girlfriend, but never spent the money to buy a baht chain to display it. More than one of our members woke up in a whorehouse bed to find his girl, his bracelet, his baht chain, and his wallet all missing. My instructor "Harry," experienced this twice.

I never heard of any Air America pilot or crewman using his bracelet for the intended purpose.

I also had never heard of any Air America pilot being captured until immediately before publication of this book, (Summer 2017). A friend sent me a copy of the February 1963 ARGOSY magazine article entitled:

"AMERICA'S FLYING SOLDIERS OF FORTUNE."

In this article, author Arturo F. Gonzales, Jr, tells how the pilot Edward N. Shore Jr. and FM, John P. McMorrow, were captured in May

11, 1961 by the Pathet Lao about 45 miles north of VTE. They were carrying correspondent Grant Wolfkill when they went down due to a tail rotor malfunction. They were held for 15 months before being released in August 1962, much thinner and a bit worse for the wear. Wolfkill wrote about this experience in his book, *Reported to be Alive,* published in 1965 with Jerry A. Rose.

I doubt that would have been the result for any Air America pilots in the late sixties or early seventies. Several Air America pilots are missing in action still.

Retraining in Bells was a quick course – two days of ground school and a few trips around the flight pattern, with a few full-down autorotations thrown in for good measure. I was once again ready to go up-country – as a re-cycled first officer.

Air America Bell XW-PFH at LS-32

PICTURE FROM THE ARCHIVES OF THE LATE LEE EMERY, WITH PERMISSION FROM HIS FAMILY.

I had my first Bell training flight on 1 June 1972 with instructor Captain J.J. McCauley at Udorn in N8312F.

5 June, 1972
XW-PFJ
FM Dunn

I fly my first flight upcountry in a Bell as copilot for Captain Marius Burke.

12 June 1972

James Rausch dies

You would think that being a helicopter pilot for Air America would be a most hazardous job, but I was amazed that during my 30 months there, only one helicopter pilot was killed. Jim Rausch was the only black helicopter pilot with Air America. He also was one of the most brilliant people I ever met, scary smart. In some ways, perhaps, he was too smart. In addition to being exceptionally intelligent, he had a remarkable capacity for language. He spoke Thai, Lao, and several of the local dialects of the area. His ease with languages may be what got him killed.

He was flying an H-34 into a non-contested LZ, northwest of Xieng Lom (*B-9 on map*). He was the one doing the talking to the Lao soldier on the ground. Normally any one of the rest of us would have had to rely on the Lao copilot to translate for us. The copilot might have realized perhaps that this LZ was not so safe as the soldier on the ground was saying.

Rausch, talking directly to the soldier, believed the LZ was safe and continued his approach into the LZ. He caught a sniper round in his right eye, dying instantly.

On the other hand, I kept a rough mental tally, and fatalities among fixed-wing crews averaged almost one a month.

17 JUNE 1972

FIVE BURGLARS ARE ARRESTED IN THE WATERGATE BUILDING IN WASHINGTON, D.C. IT WILL EVENTUALLY BE SHOWN THEY HAD TIES TO PRESIDENT NIXON.

24 June 1972

I pass the two-year mark with Air America.

7 July 1972

A good friend, Steve Howell, arrived to work with us. He brought his wife, Annie, a schoolteacher, with him. He was another Vietnam buddy whom I knew I could trust to be there for me in a crisis.

I was flying with him in Vietnam one night when our gunner opened fire over the village just outside the base. We were on final approach after a routine night medevac. The gunner was not supposed to fire without permission from Howell, the aircraft commander, but he saw some flashes and thought we were being shot at. The flashes he saw were the lights of the village as we passed over on short final approach to our base. When he opened up with his M-60 machine gun, his bullets killed a little girl asleep in her bed. Steve was in trouble until it was determined that he had indeed briefed the gunner to not fire without orders. I don't know what happened to the gunner after that.

Howell's arrival brought our total pilot complement up to about 200 pilots flying helicopters out of Udorn. Before I left a few months later, the total approached 230.

Steve's wife Annie was a bit of an artist. She created the original version of the tee shirt,"When you are out of H-34s you are out of helicopters." She used the "stick figure H-34 from the Navy/Marine Corps NATOPS manual. She sent me a pair, for my wife Michele and I. Michele said, "I can do better than that." She then created the smiley-face H-34, which has become the logo for 'popasmoke,' the Marine Corps Helicopter and Air Crew Association that has more than 2,000 members. We recently invited theV-22 Osprey pilots to be part of our association.

56

STO July 1972

Hong Kong

Now having 10 days off every month, I journeyed to HKG alone. Gary would follow in a few days. I proceeded up to the Apollo Lounge at the top of the Hyatt Hotel. It was early and things were quiet, too early to expect any Pan Am crews. There were a few pretty Chinese girls and one fellow at a large corner booth. Management thought it would be bad for the reputation of the hotel if there were an appearance that prostitutes worked the Apollo Lounge, so unaccompanied women were not allowed in the bar, hence the single male with several women. (This rule worked much in my favor the next month.) As I walked by, one of the girls smiled at me and invited me to join them. At first I hesitated, then accepted. The group was recruiting members to join a singles club called the "Up Club." I didn't feel the need to join a singles club. I was getting plenty of action on my own.

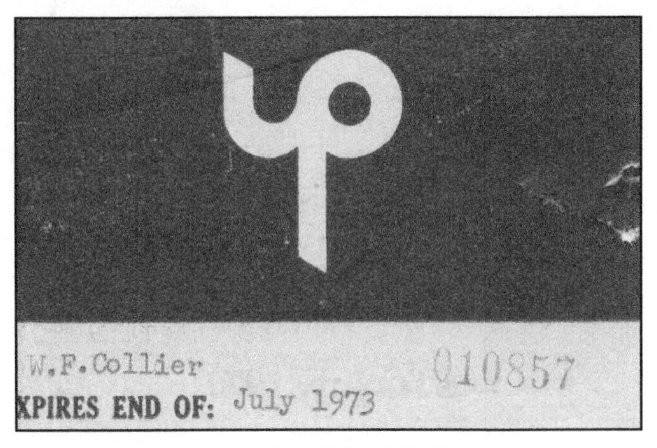

AUTHOR'S SCRAP BOOK.

After a bit of conversation, they told me that there was a party coming up – a toga party at a local health spa. Everybody would be wearing only a toga made of sheets. This began to get interesting. I objected, "I don't have a sheet and I don't feel comfortable stealing the sheet off my hotel bed." One of the cute young Chinese girls countered, saying, "Sheets will be provided." I finally caved in and paid my $15 annual dues and agreed to go to the toga party. When I showed up I could tell we were going to have a good time. There were only about a dozen people, but there were abundant drinks, food, a swimming pool and steam room. Everybody was friendly. Sexual energy permeated the room. I realized that this situation had a great potential to turn into an orgy. It did not, but I feel that if just one person had said, "Let's all get naked," we would have.

As the party broke up, I found myself in an enviable position. With our street clothes back on, I was being pursued by two equally beautiful women. The first was a stunning knockout of a Chinese-Pakistani girl named Liz with super-model looks and figure. She was dressed in a short, tight black dress, low cut and quite revealing. Unlike most Asians, she had a bit to reveal. She was almost irresistible. The other girl was an attractive redheaded, green-eyed Irish lass, dressed a bit more conservatively. Her captivating Irish accent garnered my full attention. She made no bones about letting me know that her roommate was a flight attendant and was out of town for a couple of days. I went for the sure thing, as I was not sure I would be able to get Liz past the front desk of my hotel. Liz was definitely not a prostitute, but try explaining that to hotel security. Margaret and I departed the party for her flat and enjoyed each other's company for the rest of the night.

I caught up with Liz at a later date.

57

Black Ops
Silent Helicopters

On one occasion while still flying H-34s, I was working out of Pakxe. I was told to go to the top of a small hill just north of Pakxe and "report to the customer." I landed at the small gravel airstrip PS-44 (*M-18 on map*). The customer emerged from a concrete bunker complex, and directed me to fly down to the nearby Mekong River and haul up barrels from the riverbank. At the river a crew of local laborers filled 55 gallon barrels, some with water and some with gravel from the river.

It was external load work and quite simple to latch onto three or four barrels at a time and sling them up to the small airstrip. The round trip was not five miles. I did this all day for three days. I felt like I hauled enough gravel and water for them to pave their entire small airstrip with concrete. Others, probably the Porter pilots, had flown in cement in 94-pound bags. I saw many bags stacked on pallets beside the bunker.

Curious about what in hell they were going to do with all this water and gravel and cement, I asked the customer. "Oh, we are just going to expand our bunker system here on the hill." Okay, I accepted that. I knew it was not good to be too nosey in such matters. I never got back up there, so I never got to see if they did, indeed, expand their bunker system.

Many years later, about 2010, I learned what really happened to all that material I hauled up there. The CIA had contracted with

Hughes Aircraft Company to create a super quiet H-500 helicopter (OH-6 "Loach") for a most clandestine mission. The helicopters had special rotors made to be whisper quiet. The tail rotor was modified to be almost silent. The engine was baffled in such a way to be inaudible from a short distance. This helicopter could not be heard from more than a couple of hundred feet away, and then it sounded like it was a quarter mile away. It was truly a stealth helicopter.

That helicopter was flown into North Vietnam at night and hovered over a telephone pole on top of a mountain. Here, specialist electronic techies tapped into the major communications line from NVA headquarters, somewhere near Hanoi, to the southern provinces of North Vietnam. We could then intercept much of the enemy's internal communications. This was a major coup of the war.

My little part in this? I helped transport some of the materials used to build a hangar in the deep woods where these silent helicopters were hidden away. Few people knew about this project until 2010.

Here is what Wikipedia has to say about this operation:

"The Quiet One"

"A heavily-modified pair of OH-6As were utilized by the CIA via Air America for a covert wire-tapping mission in 1972. The aircraft, dubbed 500P (penetrator) by Hughes, began as an ARPA project, codenamed "Mainstreet", in 1968. Development included test and training flights in Culver City, California and at Area 51 in 1971. In order to reduce their acoustic signature, the helicopters (N351X and N352X) received a four-blade 'scissors' style tail rotor (later incorporated into the Hughes-designed AH-64 Apache), a fifth rotor blade and reshaped rotor tips, a modified exhaust system and various performance-boosts. Deployed to a secret base in southern Laos (PS-44) in June 1972, one of the helicopters was heavily damaged during a training mission late in the summer. The re-

> maining helicopter deployed a wiretap near Vinh, Vietnam on the night of 5–6 December 1972, which provided the United States with useful information during the Linebacker II campaign and Paris Peace Talks. Shortly thereafter, the aircraft were returned to the U.S., dismantled and quietly found new homes as the now-standard 500s."

Anyone interested in more detail on this will find a link to the actual CIA file in the Wikipedia article.

The Company also acquired a pair of H-500s to use around our areas of operations, probably to provide cover for the quiet ones in case one might be spotted. "Plausible deniability." The Company also acquired a few H-47 Chinooks about this same time. They were used to support the troops on Skyline ridge and elsewhere.

13

AIR AMERICA LOG ★ แอร์อเมริกา

UDORN's LATEST CHOPPERS

Latest addition to Air America's helicopter fleet, a Boeing Vertol CH-47C, eight of which have been assigned to the Company's Udorn Base. (Pix by: Mike Kondr, OM/UTH)

Another new addition to Air America's helicopter fleet at Udorn, the Hughes H-500, taking off in front of a hangar at Udorn.

"PROFESSIONALISM THROUGHOUT"

An H-47 Chinook and a Hughes 500 "Loach."

58

About Parachutes

With the Strela threat, I evaluated the situation and decided that I would start carrying a parachute. Parachutes, along with a variety of other survival gear (Uzis, for example), were available from the gear closet just outside the door of operations. Without asking or making a big deal out of it, I began to check out a chute and carry it out to my helicopter each trip on departing Tango. I simply removed the pilot's seat back with the parachute and left it in the helicopter for the duration of each trip, usually six days. I received a little derision and scoffing from my fellow pilots, but I ignored them. I knew my chance of getting out of one of these Bells in a chute was slim, because the rotor system or tail rotor would probably eat me even if I did have time to bail out at the low altitudes where we worked. But I also knew that sometimes the Bell's reaction to a sudden shock to its airframe was to throw the entire rotor system off the aircraft. This, I reasoned, would give me a second or two to jettison the door, roll outside, and pull the "D" ring … with no rotor system to worry about.

All this presumed that I hadn't been killed by the blast of the missile, would still be conscious and alert enough to figure out instantly what had happened, and still be capable of rolling out of the aircraft and pulling the D ring. Not much of a chance, but when I visualized myself riding down out of the sky in a burning, rotor-less helicopter, I figured that the hassle of the parachute was worth the possible benefits, however miniscule.

A few months later, just after I left the area, I heard that parachutes became mandatory for all pilots in all helicopters.

A perfect example of the fragility of the Bell rotor system was provided a few weeks before I got into Bells. I returned my H-34 to VTE one night after a hard day's flying upcountry. As I walked to operations from my helicopter, I saw that fire and crash trucks had gathered, lights flashing, at the far end of the ramp. I was too tired to walk to the other end of the ramp to see what was happening. Just a hot start, I thought.

An hour later in the bar, I learned what had happened. One of the senior Bell pilots, parking in a revetment, backed in a wee bit too far and struck his tail rotor on a concrete lamppost. The helicopter dropped three to four feet to the ground, landing hard on its skids, spreading them. The shock to the drive train caused the rotor system to depart the helicopter. The entire rotor system, intact, then swirled its way up into the night air, across the tarmac and struck another Bell, (XW-PFH), which had also just landed and shut down. The copilot of that second one, Larry Hennessey, had just stepped out of his seat and still had his hand on the door handle, when the main rotor of the first neatly skewered the seat he had just seconds before vacated. A few seconds sooner and he would have been violently sliced in half. Part of the tail rotor of the crashing Huey passed through the concrete block wall and killed a pig. The tail rotor transmission was never found.

One day I almost made use of the chute in the traditional way. I was deadheading from Long Tieng to VTE as a passenger in the back of a Bell piloted by Ted Cash. Ted got up through a sucker hole on top of the clouds and then, of course, the low fuel light came on. I sat by the open cargo door and watched the engine gauges over Ted's shoulder. At the first indication of the engine quitting, I was going to roll out the door and push myself away from the helicopter with both legs to avoid the tail rotor, and yank the D ring. There was no way I wanted to ride that helicopter down through the trees. Even though I knew that the odds would be marginal for the chute to properly function, I decided I would much rather take my chances in a parachute. Fortunately, Ted found a hole in the clouds; he landed and refueled at LS-272 before proceeding to Vientiane.

14 JULY 1972

DEMOCRATS CHOOSE SENATOR GEORGE MCGOVERN OF SOUTH DAKOTA AS THEIR PRESIDENTIAL NOMINEE. HE IS AN OUTSPOKEN CRITIC OF THE WAR.

59

STO to Australia

Summer 1972

Gary and I decided to go to Sydney, Australia for our STO. Of course, we flew Pan Am. Because of my longer STO, I traveled solo each way. I met up with Gary at our hotel in King's Cross, right next to the Texas Roadhouse, which was a huge mecca for GIs on R&R. It attracted a huge crowd of young ladies who came to meet the GIs. If you couldn't have a good time there, you were not alive.

Our first night out, Gary introduced me to his partner for the night, Helen. She confided to him that she was a nanny for a rich family somewhere outside Sydney. She got only Saturdays off, and her main goal every Saturday was to come to the Roadhouse and get laid. He made sure she reached her goal that week.

Gary had contacted Claire, the piano player from the Siam Intercontinental Hotel. She set us up with her boyfriend who took an entire day off work and gave us the grand tour of Sydney. We felt honored.

Two days later, Gary phoned my room earlier than I would normally expect. He said, "Bill, is there any wine left over from the bottle we bought last night?" I was puzzled. Yes, we drank a bit but we never began drinking so early in the morning. Besides, the local white wine we had bought was awful. He said, "Bring it over." When I entered his room, he was sitting in bed beside an attractive blonde from New Zealand; both were naked. There was no mistake in this invitation. I shucked my clothes onto the floor and got in bed on the opposite side of the blonde. Things went downhill from there and we had a most intriguing and satisfying three-way frolic for about two hours. After a while we all three fell back in a pile of exhausted bodies and had a long, nervous-tension-relieving giggle fit.

Gary had to leave before me and abandoned me to my own resources. It was Saturday night again. I went next door to the Texas Roadhouse. As I walked in, I spied Helen handing her coat to the coat-check girl. I walked up to her and, with no small talk, said, "Helen, get your coat." She retrieved her coat. I took her by the arm and guided her to my hotel room so she could fulfill her goal for that week.

On the way back to BKK, I cornered yet another delightful, bright-eyed blonde, Annie, in the back galley of the 707. We hit it off, and we made a date for our layover in Tokyo. We went out to supper and retired to her room. She said she was from Estonia and spoke Estonian. I asked her to say something in her native language. She said she could do better than that, so she told me a bedtime story in Estonian. When she was finished, I told her it did not make any sense to me. She replied, "That's Okay, it doesn't make sense in Estonian either." Like Susan, she was using her flying job to work her way through college at San Diego State. I would visit her at the tail end of my 1972 annual leave.

16 July 1972

Flight Mechanic Feliciano C. Manalo is killed by gunfire while his pilot is attempting a medevac near Pakxe. I liked him, he was a good man.

18 JULY 1972

JANE FONDA BROADCASTS ANTIWAR MESSAGES WHILE SHE IS IN NORTH VIETNAM. SHE IS PHOTOGRAPHED SITTING IN A NORTH VIETNAMESE ANTI-AIRCRAFT GUN.

21 July 1972

A pilot is killed east of Pakxe. His plane ended up in a tree. No other details.

25 July 1972

A crew of three – Captain Benjamin F. Colemen, First Officer John T. Grover, and crewman Thanom Khanthaphengxay – are killed when their Twin Otter hits a hillside while trying to drop arms to troops in contact near LS-72.

"Flying, like the sea, is terribly unforgiving of human error."
Paraphrase of old aviation saying.

5 August 1972
XW-PFJ
FM Bob Noble
Copilot for Frenchy Smith

I carried U.S. Ambassador G.M. Godley to Nam Phong. No other details in my log book.

I celebrate my 29th birthday this month.

18 August 1972

On a ten-day STO, I ventured to Pattaya Beach, a resort area on the Gulf of Siam, 90 minutes east of Bangkok. I had passed the two-year mark with Air America in June. Two years down, one to go on my three-year contract. I tied up with a handsome, leggy Swedish blonde girl and enjoyed her company for several days … and nights.

27 August 1972

Captain William F. Reeves, First Officer Joel Gudahl and crewmen Thongham Khammanephet and Praves Satarakij are killed when their C-123 hits a mountain en route to Long Tieng. They carried a load of "hard rice" (explosive ordnance).

60

I Carry a Lot of Baggage

After packing the chute around for a few weeks, I realized I was packing around a lot of gear. In addition to the chute, I carried a survival vest full of survival equipment including an emergency radio. We were allowed to carry weapons for "survival purposes only." I carried an Uzi sub-machine gun with 200 rounds of 9mm ammo, a sawed-off M-1 carbine with four clips of ammo, and a Walther PK 9mm pistol which conveniently used the same ammo as the Uzi. I also packed my personal suitcase, which held several changes of uniforms and always a bottle of JW Black. My map case weighed about 10 pounds, and my Cloud Nine cushion weighed in at 16 pounds. Out of curiosity one day, I carried all my junk down the hallway to the Company quack's scale and weighed it: 62 kilos. I was carrying around 135 pounds of gear with me on every flight.

Joe Lopes discovered the Cloud 9 cushion. It was developed for paraplegics who have to spend their entire lives seated. It relieved a lot of soreness from prolonged sitting. I bought one and absolutely loved it. Before I bought it, whenever I had a long day of flying the H-34, I felt like an old man at the end of the day, with a painful back and sore legs. I had trouble straightening out when I climbed out of the cockpit and had to walk halfway bent over for the first few steps.

After I started using the Cloud 9 cushion, I could step out of the cockpit after a hard day's work and walk away like I had not been sitting at all. I loved it. I wore the first one out and bought a second. I recently tried to find that gel-filled cushion, and it seems to have been replaced by dense foam products. Too bad.

The picture on the carton this product came in showed a svelte young woman sitting on an egg to demonstrate how well the cushion wrapped around one's bottom. One night several of us were sitting around socializing. I saw the Cloud Nine box against the wall and said, "I've got to see if this is true." I took an egg from my fridge and slowly lowered myself onto the egg. Sure enough, the cushion swallowed the egg without it breaking.

28 August 1972

I was in the Udorn U.S. Air Force Base officers' club when Air Force Captain Steve Ritchie entered the bar. He had just made his fifth Mig kill over North Vietnam. He was now the first USAF pilot ace in the Vietnam War. Wearing a beige jump suit with five huge red stars sewn onto it, he was strutting his stuff!

13 September 1972
N8596W|FM Ortillo
Copilot for Broz again

This day we flew 11:40 hours, two minutes longer than my longest day in H-34s, when I almost got stuck out overnight in the Lao village. I had many days that were nearly this long in Bells. I got my 100-plus hours of monthly flight time quickly. No wonder our STOs were ten days.

61

Second Annual Leave

26 September to 4 November 1971

Gary and I enjoyed our second annual leave. We again attached our STOs to either end of our leaves. For a second year, we had six weeks off with pay. We again bought PAA Flight One tickets with every stop written in, our second epic around-the-world vacation.

By this time we had been corresponding with several of our stewie friends. We tried to set up dates for some of our stops. Our first stop in 1972 was Athens, Greece. We stayed at the Omonia Hotel on Omonia Plaza, absorbing the local ambience. This was authentic Greece, not a big, plastic American hotel like a Sheraton or a Hilton. Informed that the hotel had no single rooms, we paid for doubles, $8 a night. The first night there I brought out a bottle of red wine I had carried from Udorn. When I opened it, it had gone a bit bad. *Hell,* I thought, *I carried this bottle this far, I am going to drink it anyway.*

Con's stewie friend, Sweet Lyn, joined up with us at the Omonia. She had another PAA stewie friend arriving the next night for me – an international blind date. The first night the three of us sat on the veranda overlooking the square, drinking Retsina, the Greek resin-tainted wine, and smoking. I did not smoke, ever, but at this time I played around with acting like I was trying to smoke cigarettes. Mostly I made jokes about those who did smoke.

After Sweet Lyn's friend, Judy, showed up, we all boarded an overnight, inter-island ferryboat for Mykonos. We spent two or three days doing the tourist bit there. At one point, in a small shop, I saw a plastic 360-degree protractor. I told the girls I needed to buy this. They asked, "What for?" I explained, "I need this to help me navigate my way around as I fly all around Thailand and Laos. You see, I just lay it on the map, orient it towards north, and then I can draw a line on the map to my destination." They were appalled that I should use such a rudimentary navigation device.

Judy and I did not hit it off, so the four of us did not spend a lot of time together. The young women went one way, and Gary and I, another. We visited Lesbos, (pictured below) and then returned to Athens, where we caught a plane for Rome. Gary and Lyn made plans to rendezvous later in Madrid. She said she had "other business" to take care of in the interim.

Gary (top) and me on Lesbos Island. We were intrigued by the story that in the ancient past this island was said to be inhabited solely by women.
AUTHOR'S PHOTO ALBUM.

Trattorias

In Rome, we took an all-day bus tour that hit all the usual tourist high spots. At least now we could say we saw a bit of Rome. (I did not see more of Rome until 2014.) Lyn had told us that the best food in Rome was at the local diners called "trattorias." Gary and I checked into our hotel and then walked down the street looking for one of these trattorias. We were hungry after our arduous trip from Athens.

It was a bit late for lunch, so when we finally got settled into one of the basement eateries, we were the only customers. On the menu, there were only about ten selections, so we made a plan. Gary would order no. 1 and no. 2, and I would order no. 3 and no. 4, and we would share, family style. The servings were smallish, so at the end of those four items, we ordered nos. 5 and 6, and nos. 7 and 8. Those arrived, and we ate them. About this time, I noticed some of the kitchen help peeking out from the kitchen doors, looking at these two gringos, eating so much.

After dishes no. 7 and 8, we decided we needed more food, so we ordered nos. 9 and 10, and to round things out, each of us ordered again whichever thing was our favorite from the first ten. I remember I had another bowl of the minestrone soup. All the time we had been drinking wine with our food.

Now the kitchen help was calling to their friends in the back of the restaurant and telling them to look at the two gringos eating so much. (I asked a friend who is a fluent, native-Italian speaker, what is the Italian equivalent of "gringo." She could not help me.) That did not bother us. Finished with our main meal, we ordered a cheese and fruit plate for dessert. Oh, and a second bottle of wine, of course. As I recall, the total bill for this feast was about 10 or 12 dollars.

We left the eatery feeling stuffed. No one was about; it was siesta time. All the Italians were home sleeping off their lunches or making baby Italians. Walking down the street, we paused in front of a storefront and were admiring all the items in the store window. We looked at each other, laughed and shook our heads when we realized we were now standing in front of a deli, admiring all the delicious hams, cheeses, and other foods in the window.

We hopped over to Madrid. Lyn had previously arranged to meet us at a particular bar that was known to be a rendezvous point for American expats. We arrived in Madrid a day earlier than Gary was to meet up with Lyn, so we decided to go to the bar to check it out. Across the smoky bar, we spotted Lyn with another fellow. To not embarrass her and cause any problems between her and that fellow, we downed our drinks and slipped quietly out. We never told Lyn we saw her with another man. In those days, we were all being quite promiscuous, so it did not bother us to know that one of our sweeties was also sleeping around. With our respective lifestyles, it was assumed.

All the time Gary and Lyn dated, she kept telling us that when we got to Spain we had to have paella, the Spanish national dish. We asked her to tell us what was in it, and she always bogged down after saying it came on a huge platter and had a lot of fish in it. It sounded wonderful, so the next night after we rendezvoused with Lyn, we went to a prominent nightclub, sat near the stage to get a good view of the flamenco dancers, and ordered paella for three. Lyn just kept raving how wonderful and different this meal would be, and how much we would enjoy it. After what seemed an interminable delay, our huge steaming platter of paella was placed on our table. Gary and I looked at it and again shook our heads. We had seen this many times before in Asia.

To us it was just another heaping platter of fish-fried rice.

A pan full of Spanish paella (WIKIPEDIA)

We bought tickets and attended a bullfight. Before we entered the arena, we stopped at a tiny boutique wine shop where the shop owner filled our bota bags from large casks right behind the bar. I amused the others when I chose my wine by asking for a red wine that would match my shirt. I knew I would be dribbling some.

As we sat cheering for the bulls, we made the acquaintance of another couple behind us. The young lady of the pair came on to me pretty aggressively, to the point where I felt uncomfortable talking to her in front of her companion. She made it a point to let me know that it was Okay for me to pay attention to her. She told us that they both belonged to an international sex club, and that members got together for sex only. Nobody cared that their partners might be seeing someone else before or after an encounter.

We made a date for the next night. We enjoyed a nice meal at a Russian restaurant and then adjourned to her apartment for a romp. I never took advantage of getting involved in the international sex club. It might have been most interesting, but I was getting plenty of action without getting formal about things. I have always wondered how that might have worked out, having easy sexual access to hundreds, perhaps thousands, of young European women. The world was our sexual playground.

Gary and Lyn stayed in Madrid, and then they skipped over to Lisbon, Portugal. Without a date, I felt I was an awkward third wheel, so I skipped Portugal and hopped the next plane for Stockholm, Sweden. There I caught up with my leggy Swedish blonde friend, Christine. She showed me around Stockholm.

I then popped over to London and spent a few days with Syn. She had changed her Pan Am base from San Francisco to London and was studying Russian. I think she may have developed a serious interest in a Russian fellow. That did not seem to change my welcome to her flat. We went to the theater and saw the play Godspell.

Thence on to Miami. I had a good friend, Dave Inman, whom I wanted to visit in Fort Lauderdale. We got in some good scuba diving and fishing off the coast. Dave and friends created the company U.S.

Divers. He retired a wealthy man.

Gary and I reunited in Sacramento, our home area. We met each other's friends and partied a bit. Before departing California, I visited Micky in Sacramento, traveled to San Francisco to visit Trans International Airlines stewie Dotty, to Oakland to visit TIA stewie Cindy, and to San Diego to visit Pan Am stewie Annie, the sweetie I met on the flight from Sydney to Hong Kong. Then I jumped back on to TWA flight 745 for the long flight across the Pacific. I had to get back to Udorn to get back to work. I needed some rest.

"I spent all my money on whiskey and women. The rest of it I wasted."
W.C. Fields

62

Working Bells at Long Tieng

5 November 1972
XW-PFH
FM Dimaandal, J.
Copilot for Vlad Broz

My first day back from annual leave

We were flying resupply to the Charlie Pads above Long Tieng. We heard a Mayday call: a Raven spotter plane was going down on the PDJ. We turned and raced towards the situation. I got out my Uzi, locked and loaded it and fired a test burst out my window. It worked as advertised. I left the safety off, ready for battle. Immediately the mission was cancelled. Someone beat us to the rescue. We went back to routine work resupplying the various Charlie Pads.

This may be the mission that is causing confusion between my recollections of this rescue on 7 November and another one, just two days later. I never knew any details of what happened out there on the Plain of Jars this date, who did or did not get rescued, who got shot at or shot up, if anyone.

Walking across the ramp one day, I watched as a TwinPac flew by with a cargo net full of barrels of water. One of the barrels fell out of the net and crashed upon rocks uphill from the ramp. The barrel ruptured and the water shot out through a crack with such force that it cut down small trees.

The Thai and Lao troops celebrated some of their most important ethnic holidays by sacrificing a water buffalo. It was important to them that the water buffalo be delivered to them alive, to be slaughtered on site as a part of their ceremony. Having a dead buffalo brought out to them defeated their purpose. This presented a problem, as it was difficult to transport a live buffalo in any of our aircraft. Finally, it was decided that the water buffalo would be placed on cargo nets, then shot with a tranquilizer gun to make them passive. Strapped into the nets, each was transported to a hilltop LZ as an external load underneath one of our helicopters. We carried many water buffalo out to the troops (See cover). There is more than one story of one of the huge beasts coming out of its daze to find itself flying through the sky, petrified and bellowing with fear.

I believe at least one of the huge beasts was dropped by accident from several hundred feet. The story was that the troops on the ground near where the animal fell were grateful to have their sacrificial beast, but were a bit miffed that we did not deliver it to them live.

Here is a picture from the Air America Log, Vol. VI, No. 1, 1972, of one of our Bells hauling a donkey. The publication gives no other details.

Air America Bell Huey "Hauling Ass"

Another aerial view of the secret CIA airbase at Long Tieng, looking eastward.
PICTURE COURTESY OF PETER AND BAYSONG WHITTLESEY.

This secret base, just south of the PDJ, was in a bowl. The short 2200-foot runway sloped upwards from the east to the west. Airplane pilots had to descend into the bowl and then make an approach to the runway that required a bit of a climb. They dare not take up too much runway on landing because there was a karst at the west end. There was no going around. Company pilots routinely flew C-130s, C-123s, DC-6s, DC-3s and smaller fixed wing into this base.

I saw the aftermath of a Lao T-28 that landed long and hot. The pilot managed to turn the T-28 right into the helicopter ramp. He missed the karst, but ran into a concrete building. The aircraft penetrated the block building and was destroyed. The pilot died. Another Lao T-28 pilot did the same thing and ran into a parked C-123, killing a local passenger on board the C-123.

The gutsy Lao Air Force T-28 pilots were committed to helping their country fight the communists. Their enlistment had no termination date. Many died.

63

Another STO to Hong Kong

Submariners At the Hilton Bar

Summer 1972

On the prowl, I took the elevator up to the Apollo Lounge. I frequented it enough that I knew the hostess by her first name, Kip, and even gave some thought to dating the little Nepalese cutie. I knew that the Pan American stewardesses were due to appear shortly. Their flight should have landed an hour ago. They always went to the bar after checking into their rooms and freshening up a bit. No Pan Am stewies in sight tonight. Their flight must have been delayed. No one from the Up Club either. Drat.

This was going to be a quiet, lonely night. I departed the lounge, bounding down the stairs when I saw Kip talking to two beautiful young ladies, in their early twenties. She was explaining to the youngsters that unaccompanied women were not allowed into the bar. The ladies could not comprehend that Kip might think of them as prostitutes and were arguing with her to gain access to the bar.

I instantly assessed the situation and I said to Kip, "Kip, these young ladies are with me. I have been waiting for them." Then I turned to the young ladies and said, "Where have you been? You're late! I had given up on you, and I was just leaving. Let's have a drink." I don't know that Kip actually believed that I had been waiting for these girls, but she let them join me. We adjourned to the bar and began to get acquainted. Gary was coming to join me in HKG the next day. I now had dates for both of us.

We often went to the San Francisco Steak House, which was right outside the door of the Hyatt. We did not often get real food where we worked, so it was a nice touch for us to get American steaks, fresh salads and such. We always ordered wine, too. It was a bit expensive, but we could afford the place. I often made this toast: "To good food and good wine and good times, all made better by the presence of good friends!"

I always have been an excellent speller. In our small studio apartments, Gary and I would often be in our respective bedrooms writing letters. Sometimes he would ask me through the paper-thin walls in a normal conversational voice, how to spell some word. I would tell him. At some of the dinners, he would brag about how well I could spell and ask our companions to challenge me to spell any word. Frequently, people would ask me to spell "rendezvous". That was too easy. Then someone asked if I could spell a word alphabetically. A sample of this would be rendezvous: d-e-e-n-o-r-s-u-v-z. When we tired of that game, I would spell words alphabetically backwards: z-v-u-r-s-o-n-e-e-d. I won a few drinks doing that. For me it was simple. I just visualized the word in my mind and picked out the letters in sequence.

After picking up the two young ladies, I took them to the Den, the nightclub in the basement of the Hilton Hotel on Hong Kong Island. It was a classy place and drinks were expensive. My intention was to get to know them better, choose one and separate her from the other. We sat at a nice table at the edge of the bar near the window overlooking the pool.

At a table next to us sat an elegantly dressed French couple out for a night on the town. Five young men occupied a third table near us. It soon became obvious to me that they were young U.S. naval officers from a submarine in the harbor. They were pretty drunk and getting more boisterous by the minute. One of them made a feeble effort to join my table and snake one of the ladies away from me, but he did not go so far as to join us. They had a higher priority at the time. They were celebrating the fact that one of them had that very day received his golden Dolphin wings and was now a qualified submariner. It was a big deal to them and they intended to celebrate it in their customary manner. (I could relate. I remembered the joy of receiving my golden Naval Aviator's wings in December of 1965.)

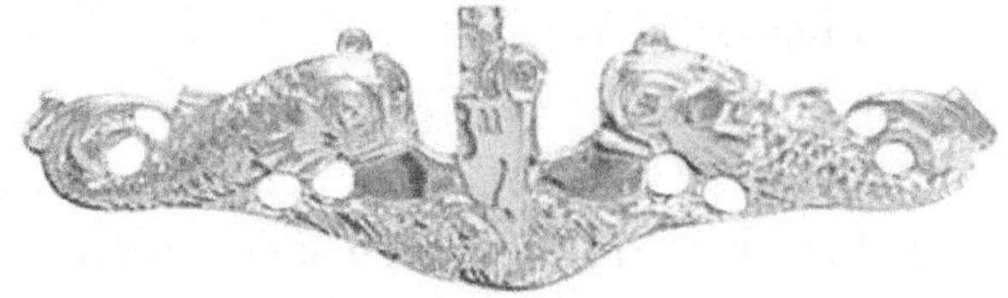

Submariner's Dolphin wings. (WIKIPEDIA)

Their celebration procedure usually culminated with throwing the newest member of their elite group into the ocean or bay wherever they were anchored. They bantered about the idea of hauling the new guy out to the Hong Kong harbor and throwing him off the pier, but nixed that idea immediately. Hong Kong harbor was known to be one of the nastiest waters in the world. They wanted to initiate him, not infect him with a dozen communicable diseases.

They discussed throwing him into the hotel pool, which was in plain sight right outside the windows. One of the fellows went out to the pool, and returned with the bad news that the pool was locked up for the night. Drat! No dunking there. I overheard them talking about taking the victim up to a hotel room and submerging him in a bathtub full of water, and they even jokingly discussed stuffing him down the toilet, but no one would buy these meager offers.

In one of those magical moments when everybody clicked on the same frequency without anyone verbalizing it, the other four submariners decided they would dump their beers onto the initiate. They all stood up, grabbed their full beer mugs and slung the contents towards the intended victim. The inductee must also have been tuned into the mental telepathy and knew the beer was headed his way. Just as the beer took flight towards him, he ducked out of the way of the in-coming beer-fall. Four full mugs of beer flew past him, splashing upon and soaking the elegant French couple. All hell broke loose as the Frenchman exploded in a rage. Think of a cigarette pack-sized nuclear device going off in an enclosed space. It was not a pretty sight. I am sure this incident added to the negative impression of "Ugly Americans." The submariners were asked to leave the bar; they left peacefully.

After things quieted down, I cut one of the sweeties (Janet) away from the other. Janet was medium height, had soft brown hair and hazel eyes, and an effervescent personality. She had just completed a mid-west college and was on a tour with her parents and brother. The two of us sat in the coffee shop in the Hyatt Hotel all night getting to know each other. My good friend, Gary, showed up the next day and entertained the other young lady. We double dated for the five or six days that our paths coincided in Hong Kong. Janet and I corresponded for months after that, and we dated after I returned to the San Francisco area.

Meeting pilots inspired Janet to become a flight attendant. She flew for Trans International Airlines for years afterwards. It was a trick of fate that kept us apart a few years later. This beautiful young lady is still a dear friend. Last I heard, the other girl, petite Celine, moved to Japan and married a Japanese businessman.

Later that week Gary and I attended an Up Club function, a spaghetti-eating contest at a local restaurant. Both Gary and I considered ourselves big eaters and were certain that one of us could win this contest, no challenge. Below is the picture of us losing the contest heartily to a skinny little Italian fellow. The Italian and I got our pictures in the Hong Kong Post Herald.

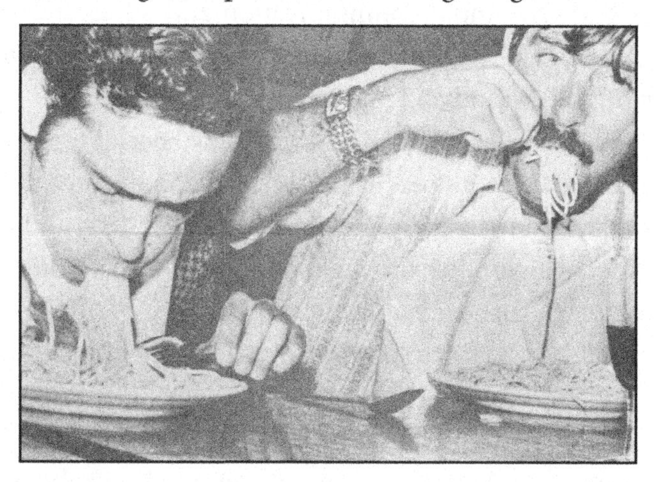

Spaghetti eating contest in Hong Kong. Notice my gold bracelet on my arm.

64

My Scariest Mission Ever!

7 November 1972
XW-PFH
FM Neufeld
Copilot for Ted Cash

I was again flying with Ted Cash and FM Neufeld. We heard a radio call that Raven 2-0 had been shot full of holes over the Plain of Jars. His engine shot out, he was gliding down in an area known to be infested with North Vietnamese Army soldiers. We were already on the top of Skyline Ridge resupplying troops, so we were just a few miles away. There was no discussion. We immediately headed north to assist.

We had clear radio communication and we carried on a conversation with the pilot as he descended. We heard him report that he was safely on the ground and had evacuated his airplane.

In less than ten minutes after his Mayday call for help, we arrived overhead. We could see his tiny airplane on the ground. I locked and loaded my Uzi machine gun, clicked off the safety, and held it up ready to shoot at anybody who might be a threat. I was willing to shoot through the Plexiglas nose of our machine if need be.

We made a cautious spiraling descent, watching for enemy activity. We fully expected to receive heavy fire during our descent. Since we had been talking to him after he had exited his plane, we assumed the Raven pilot would quickly scamper over and jump aboard our helicopter. We would be off in an instant. Ted hovered in close to the tiny aircraft to facilitate the pickup. Sitting in the left seat, I could not see anything that was happening on the right side of the Bell. Ted and Neufeld could see the pilot lying on the ground. Part of his head was blown away. Dead! We did not know whether this was the work of a simple rice farmer shooting a single-shot blunderbuss or a battalion of PAVN soldiers. We did know the potential was enormous that there was at least a battalion of enemy soldiers in our immediate vicinity.

Ted instantly decided that there was too much risk in trying to retrieve the pilot's body. He pulled pitch (power), and we departed the area as fast as we could climb. I often wonder why we did not get hosed by a thousand bullets from the multitude of enemy troops that had to be nearby.

This was by far the scariest mission I ever flew, either in Vietnam or Laos. We knew we were descending into a hornet's nest. We knew the enemy was right there beneath us. They had just shot the Raven pilot in the head. Why the enemy did not shoot us full of a million holes, I will never know. I still get nervous and sweaty every time I reread this story or think about this event.

(The Raven pilot's body was repatriated: 04/09/2007)

That's how I remember it.

James Parker, (customer "Mule") in his books, *Covert Ops, Codename Mule*, and *Battle for Skyline Ridge*, goes into much more detail about this Raven rescue attempt. He states that there were several helicopters involved, jet fighter-bombers providing cover, and every helicopter returned to long Tieng full of holes, some needing extensive repairs, but no one was wounded

There is even a letter of appreciation written by the U.S. Embassy Air Attache Lt. Col. Hayden C. Curry, USAF, listing all the helicopters and crews involved. (See below) I cannot reconcile the contrast between my memory and what Mule reports. I have a very good memory. I know Cash and I were the first on station over the dead Raven. I don't know why anybody else would bother to descend after we reported seeing the pilot dead. I know we did not take any fire, as much as we should have. We definitely did not take any hits.

The letter of commendation below shows that I flew with Vladimir Broz whom I contacted. He denies any involvement in this piece of action, even though his name is on Air Attaché Curry's report. There was another mission on 9 November involving the rescue of a Lao pilot, also mentioned in the letter below. I truly feel two or more missions got confused on some bureaucrat's desk in VTE or even further up the chain of command.

I have also read conjecture that the Raven may have chosen to commit suicide rather than risk the lives of all the pilots and crew coming to his rescue and to avoid becoming a prisoner of war.

"I was [radio] witness to Captain Collier's death-defying effort to rescue an Air Force Raven shot down on the Plain of Jars. He displayed Medal-of-Honor guts. Collier writes a well-crafted, informative, colorful and detailed book about his involvement with the CIA's secret war in Laos."

- CIA ground officer
James "Mule" Parker

COMMENDATION

EMBASSY OF THE
UNITED STATES OF AMERICA

APO San Francisco 96352

OFFICE OF THE AIR ATTACHE 13 November 1972

SUBJECT: Letter of Appreciation

TO: Mr. James A. Cunningham, Jr.
Vice President, Laos
Air America, Inc.

1. Air America has established a tradition of heroism and bravery in helicopter operations, especially in the recovery of downed crew members. For some time now I have intended to convey my admiration for a job well done. Recently two search and rescue efforts were undertaken by your crews that were so noteworthy that I can no longer refrain from expressing my gratitude.

2. These two missions, one of 7 November and the other on 9 November, are typical of the devotion and selfless dedication of your pilots. The SAR on 7 November, although not resulting in the recovery of one of our FAC's, was so courageously supported by your company, that it has warranted the respect of all my personnel. Seven of your crews participated in this endeavor. Please extend my personal thanks to the crews of PJF, PFH, PHD, 96W, 35F and 12F. The latter two crews performed heroically and with complete disregard for their personal safety in an attempt to save a downed airman. The other crews, although not as intimately involved, were there willing and able to assist if needed. It is reassuring to all of us to know that in the case of an emergency that a concerted rescue effort will be made. This point was dramatically proven on 9 November when a downed Lao pilot was rescued within minutes of extraction by the crew of PFJ.

3. Because of these actions and of those of the past, I extend to you and your crews my most heartfelt appreciation, gratitude and admiration.

HAYDEN C. CURRY, Colonel, USAF
Air Attache

THE RESCUERS

Here are the names, by aircraft, of the Air America crewmembers involved in the SAR (Search And Rescue) missions on 7 November, 1972:

XM—PFJ	N8535F	N1196W
Capt. F. N. Smith	Capt. M. Jarina	Capt. P. L. Colgan
F/O A. W. Wilbur	F/O G. Taylor	F/O R. H. Wright
F/M W. J. Parker	F/M T. W. Yourglich	F/M J. E. Israel
F/M B. Boonreung		

XM—PHD	XM—PFH	N8512F
Capt. W. Hutchison	Capt. V. R. Broz	Capt. T. R. Cash
F/O P. G. Gregoire	F/O W. F. Collier	F/O R. A. Heibel
F/M M. A. Leveriza	F/M J. G. Dimaandal	F/M G. R. Neufeld

The rescue of the downed pilot on 9 November, 1972, involved the following crew in XM—PFJ:

Capt. J.D. Fonburg
F/O R.R. Zappardino
F/M T.W. Yourglich

(NOTE: All the above-mentioned choppers and flight crews are based at Udorn, Thailand.)

I think this old Pogo cartoon somewhat explains the mystery of what really happened 7 November 1972. With my report, we now have two reports in "black and white" that contradict each other:

Contrary to rumors, we did not receive a huge bonus for rescuing a downed airman. We did it because we were driven out of love for our fellow airmen. We knew that if the situation were reversed, the other guy would be there for us. We did it for the satisfaction of helping another pilot.

7 NOVEMBER 1972

PRESIDENT NIXON RE-ELECTED BY BIGGEST LANDSLIDE IN HISTORY TO DATE

13 November 1972
Bell N8535F
FM Neufeld
Copilot for Cash again

We carried General Vang Pao around Long Tieng area.

65

Laced at Long Tieng

Customer "Hambone" Wounded

14 November 72
BELL N8535F
FM Neufeld
Copilot for Captain Ted Cash once again

Only a week after the failed attempt to rescue the shot-down Raven, I was again flying copilot for Ted Cash in Bell N8535F. Late in the day, the customer asked us to go to the southeast edge of the Plain of Jars and look for his Thai troops. They had abandoned their position on the Plain of Jars and were fleeing from the North Vietnamese Army. I was apprehensive. Flying low and slow over the trees at dusk was asking for trouble. From experience I knew that the enemy could and would shoot at aircraft with impunity late in the day. We would have trouble getting air support to back us up after dusk. Ted asked FM Neufeld and me if we were willing to go out and look for the troops. I expressed some concern, but Neufeld did not. Ted overruled my worry. We departed northeast out of the Long Tieng bowl, over Skyline Ridge, and flew the short distance to the southeast corner of the Plain of Jars. Customer Hambone thought we should find his troops there. We descended and started circling slow and low over the trees.

I squirmed in my seat with discomfort.

My fears were proven true. An old aviation adage comes to mind:

"Flying is hours and hours of boredom interrupted on rare occasion by a few seconds of sheer terror."

We took a burst of AK-47 automatic rifle bullets into the bottom of our Bell, through the fuel cells. The bullets hit no critical parts of the helicopter and there was no fire. We remained airborne. Customer, "Hambone" in the back, whose troops we were trying to locate – the fellow responsible for our being out there in the first place – caught a round in his shinbone. It blew out a chunk of his shinbone about three inches long. He was crying out in pain. Looking back into the passenger area, I saw a puddle of blood the size and shape of an inverted dinner plate on the floor of the helicopter.

We darted back the short distance to Long Tieng. On inspection we were relieved to learn that the helicopter had no serious damage. As I started to climb out of the helicopter, I noticed that the green overhead Plexiglas right over my head had a big piece blown out of it. It was big enough to poke my head out and look around should I want to. I was shocked.

I inspected every aspect of the cockpit. There were no bullet holes in the windshield of the Huey. There were no bullet holes in my side window or door. I even patted myself down to make sure I did not have any holes in me where a bullet might have passed. No holes. No blood. How in hell did the green window directly over my head get this big hole blown out of it without my noticing?

Then I reached back between the seats to gather my pilot's kneeboard off the floor between the pilots' seats, I saw it had a bullet hole through it. The floor beneath it had a bullet hole. I roughly connected the dots between the hole in my kneeboard and the floor. The line of sight led me to the back of my pilot's seat.

I found a bullet hole in the back of my seat at chest level. The bullet core, after passing through the aircraft and the kneeboard, where it shed its jacket, struck the back of my seat. After passing through the seat, it encountered the armor plating and skittered up between the seat and the armor plating. It deflected off the upper seat frame, forward and up through the greenhouse window. It must have passed within an inch of my head. Had I not had an armor-plated seat, the bullet core would have hit me right at my heart level. The bullet would probably have hit my ribs and been deflected. By then it had expended a lot of its energy and shed its jacket, but there is no way to know for sure. It could have deflected inwardly between my ribs, into my heart. Thank God for armored seats. I still have the kneeboard and the bullet jacket that I extracted from the kneeboard.

My kneeboard damaged by AK-47 bullet with the bullet jacket that stuck in it.

Three of the times I recently flew with Ted Cash, he had almost got me killed. The first was a few months earlier when he got stuck on top of the clouds near LS-272 with the low fuel situation. The second was on the failed Raven rescue, and the third was getting us shot up with Hambone aboard N8535F.

Needless to say, I began to get a little nervous.

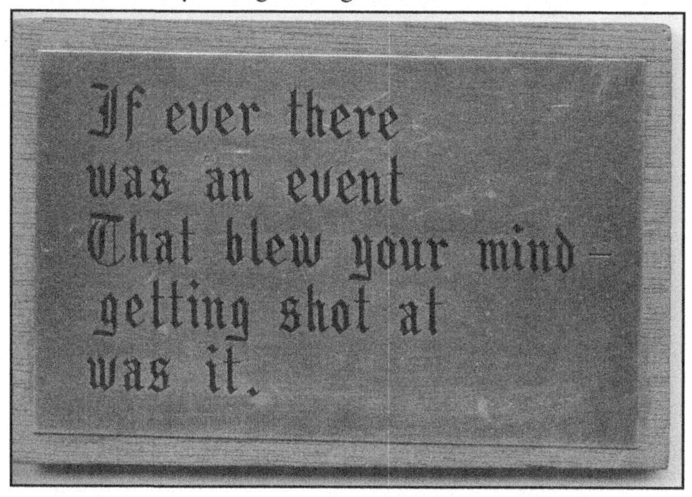

"If ever there was an event that blew your mind, getting shot at was it." Gary's family had his plaque on the wall of their Lake Tahoe cabin.

Ted Cash died in 2003 of natural causes.

66

The Beginning of the End

Last STO

20-28 November of 1972

I took what was to be my last STO and bussed 90 minutes east from Bangkok over to the beach resort town of Pattaya Beach. This was a budding resort beach area and was a great place to rest and reflect. I was beginning to ponder that perhaps my time at Air America was coming to a close. Things had been getting hotter and hotter. I had been shot up while flying with Cash. By all rights we should have been killed while failing to rescue the downed Raven. The Strela missile was reported to be in our area. Awful things were happening all over Laos. I sat around the hotel pool drinking beer and wrote in my journal for hours, trying to make some sense of the chaos around me.

To help me cope with my stress I latched onto a beautiful young tourist from South Africa named Petra. Her delightful accent along with her ample blonde hair, big blue eyes, and supple body helped me ignore my woes for a day or three.

Follow up:

13 Feb 2016

Reading over my journal recently to see if I could shed any

light on my mental state at the time, I found that the bottom line summation of my journaling seemed to be: "What the fuck am I doing here." And "Is all this really necessary?"

30 NOVEMBER 1972

AMERICAN TROOP WITHDRAWAL (FROM VIETNAM) IS COMPLETE.16,000 "ADVISORS" AND ADMINISTRATORS REMAIN.

67

Lightning Love

"An old pilot is one who can remember when flying was dangerous and sex was safe."

-Author unknown

The last night I stayed over at LPB, two nurses from the International Red Cross joined us at the gazebo. I had a long chat with one of them and found her quite charming. She was a smallish woman, about five feet tall, well proportioned, about 25 with short reddish-brown hair and honey-amber eyes. She spoke perfect English. After a few drinks, she asked me if I had ever seen the Royal Stairs, a long flight of stairs ascending from the Mekong River up the riverbank to the Royal Palace, only a few blocks away. I replied I had not seen them. We jumped into her shabby Red Cross Jeep and drove over to the stairs. We walked down a few levels and sat there for some time, having quite a lovely conversation.

After about an hour, a giant thunderstorm pounced upon us. It began to rain hard. We scampered back to the Jeep. It was a magical, exciting thing, to be isolated by the raging storm in the door-less, cloth-topped Jeep. In a flash of lightening our eyes met, communicating what our bodies needed to express. I reached over and touched the left side of her face. She nuzzled my hand. I gently placed my hand behind her neck, pulled her to me and kissed her. She responded with hungry kisses. We began to devour one another. Soon she climbed over to my side and straddled me. I was delighted to find out she was not wearing panties. We knocked off a quickie right there in the front seat of her Jeep, surrounded by the lightening, thunder, and driving rain of the pounding thunderstorm.

At one point, I rearing back, I broke the back support of her passenger seat. I have often wondered how she explained that to her fellow Red Cross workers. Sadly, I left town the next day and never returned to LPB. I never saw her again.

The number one song on December 2nd 1972
"Papa Was a Rollin' Stone"
by The Temptations

Gary told me this story and I am sad I was not there to witness it. He bought one of the very first digital calculators produced by Texas Instruments. He paid $229 for this new digital wonder, (something you get free in your cell phone these days) so he could constantly re-calculate the value of his investments. He also bought one of the very first digital watches to come on the market. He was proud of his hot new gizmos.

One night he drove over to FM Charlie Brigham's house for a few drinks with some of the fellows. Gary delighted in showing off his new watch to the guys. Someone looked at it and said, "Hey, this thing is not working." Gary was shocked and upset. His new wonder, which was supposed to be state-of-the-art, never lose a second in thousands of years, and run almost forever on one small battery, had stopped working. Knowing him as well as I did, I know he was quite dismayed.

Charlie said to Gary, "Here, let me see that thing. I can fix it." Charlie was one of our best flight mechanics. If anybody could fix anything, Charlie was the guy. Gary surrendered the fancy new watch to him. Charlie laid it in on the bar, and quicker than Gary could react, Charlie brought up a big hammer from behind the bar and smashed the watch to bits. He handed the smashed watch back to the shocked Gary.

I wish I had been there to see the look on Gary's face. The shock value was worth the joke to Charlie. The next day Charlie went to the Base Exchange and bought Gary a replacement watch.

7 DECEMBER 1972

APOLLO 17 LAUNCHED, MADE 75 ORBITS OF THE MOON. ASTRONAUTS CERNAN, EVANS AND SCHMITT ABOARD FOR THE LAST MOON FLIGHT. CERNAN IS THE LAST HUMAN TO WALK ON THE MOON. (CERNAN DIED JANUARY 16TH, 2017)

68

Mak Mak Danger!

(*Mak Mak* is Thai for *Mucho!*)

I Resign

7 December 1972

On this date my fear/greed ratio teetered terribly out of balance when Continental Air Services Captain Kemp reported a Strela firing up country. No one was hit. In fact it seemed that no one was even shot at. It was just a report of the sighting of a Strela missile's telltale corkscrew track. I happened to be at home the day it happened, and I often have wondered if things might have been different I had been working that day. If I had been up country, and the target of the Strela had not been me, I might well have said, "Oh well, it wasn't me," as almost all of the other pilots did, and kept on flying.

> *"I'm alright when I am at the front, but it's when I am back and start thinking and visualizing that it almost overwhelms me."*
>
> -Ernie Pile, famed World War II correspondent.

I immediately extrapolated that these things were now going

to be all over our area of operations. I wanted nothing more to do with this mercenary flying business, patriotic or not. The next day when the microbus came for me to take me to work, I told the driver that I was sick, and that he should go pick up the reserve pilot. I was immediately summoned to the chief pilot's office. He ordered me to report to the company quack for a physical. Upon reporting to the quack, I told him I had a bad case of diarrhea and couldn't go flying. He told me to go home and return with a sample. Instead, I went home and wrote a letter of resignation, including a request that I not be required to fly for the next 30 days.

"Did you ever wonder why
There are old pilots and
There are bold pilots,
but
There are no old, bold pilots?"

(Old aviation saying)

Because I had requested to not fly anymore, the Company fired me. I was fired for quitting. I went to the Company travel office and began to make arrangements to get my personal stuff shipped home and to get my repatriation travel. Both of these expenses would be covered by the Company because I had been fired. As soon as the paperwork began to flow, the Company realized that it was going to cost it several thousand dollars to fire me. They cancelled my being fired and accepted my resignation.

Now I was unfired.

I couldn't go home right away because I had to work off my 30 days' notice. Since I had requested not to fly anymore, the the company had nothing for me to do.

Paul Gregoire, the only other helicopter pilot to actually quit over the Strela scare, and I teamed up and began to terrorize the company bar. We got drunk every night until the bar closed. We got a bit carried away, because, well, "What are they going to do to us, fire us?"

They already did that, and then undid it. Not likely they would try that again. We haunted the bar, drinking a prodigious amount of booze in a couple of weeks.

One evening we were having an especially good time and taking turns buying drinks for all of our friends in the bar. Every time we bought a round, the bartender rang the brass bell at the end of the bar signifying that the next round of drinks was on us. We soon made a game out of throwing our empty beer cans at the bell, trying to ring it. This was our way of signaling both that we needed yet another beer for ourselves and the barkeep should set drinks up for everyone else, too.

In the corner under the bell stood a pyramid of stacked beer cans to show which types of beers were available. Soon one of us had the good luck to hit the pyramid of beer cans, which tumbled the whole stack. This was great sport. What we didn't know was that the entire stack of display was cans full of beer. When they tumbled down, one of them popped open. As luck would have it, warm beer spewed the entire length of the bar, spraying several of our friends and their wives.

As the bartender ejected us from the bar, we resisted. We were having too much fun. Finally he bribed us with a free drink if we would, "Please just leave the bar." As I was leaving the bar in good humor, I finished my cocktail and tossed the glass over my shoulder into the bushes. It did not break.

The next morning, we were both summoned to the vice president's office. Mr. Ford handed each of us a letter of reprimand about our frolics of the night before. "Read this," he said. When I handed the letter back to Mr. Ford, he asked me what I had to say for myself. "Nothing, sir, except that we didn't throw any **full** beer cans." "I appreciate your honesty," he said, "but I am going to have to restrict you from the club." "Okay," we said. Being still rebellious, we turned around, walked out the door, down the stairs and directly to the bar and ordered drinks.

"What are they going to do to us, fire us?"

Fortunately, the club manager was a nice fellow and we liked him. He explained to us that if we stayed in the bar, he would get fired. Since we did not want to see him fired, we turned around and exited the bar. Forever.

In that same bar we had played a few thousand games of darts, some serious, most just for fun. Some of the guys got pretty good, and there were times when they would gamble for $50 a game. I watched a lot of money change hands in that bar over games of darts. I usually played only, "loser buys the beer."

We had a rule that if a dart struck something hard and bounced off the board, you could throw it again only if you caught it before it hit the floor. It seemed like a fantasy, but I actually saw Sandy Sandt do it one night.

A serious game between Sandy Sandt and Joe Kivenas came to a bitter, sudden-death end. Both players had all their marks on the board save one; the next person to score would take $50 from the other. Sandy needed only a triple twenty to win, and Joe needed but a single bull's eye to wrap it up. Then they both choked up. For several rounds, neither of these drunken players could barely hit the wall, much less their tiny targets. Several times they threw, and several times they both missed their marks. It became ludicrous that neither of these ace players could score.

The bar got quiet as everybody stopped talking and watched this suspenseful sudden-death play-off. Sandy stepped up to the line and threw his darts. The first two he wasted. His third dart hit a metal ring and bounced up and away. Sandy reacted instantly, throwing himself onto the floor, sliding towards the dartboard. Whipping his hand out in front of him, palm up, Sandy allowed the falling dart to stick into his palm. He then stepped back to the throw line while coolly pulling the dart out of his hand, took careful aim, and skewered his dart right into the triple twenty that he needed to win.

He got a huge round of applause... and won $50.00.

From December 16 until December 30, 1972
the number one song on the charts is
"Me and Mrs. Jones"
by Billy Paul

18 TO 29 DECEMBER 1972

OPERATION LINEBACKER II. INCESSANT "CHRISTMAS BOMBINGS" OF NORTH VIETNAM. BOMBERS TOOK OFF FROM UDORN EVERY 30 SECONDS FOR THIS ENTIRE PERIOD. DURING THIS OPERATION MORE THAN 100,000 BOMBS WERE DROPPED ON NORTH VIETNAM. FIFTEEN B-52S WERE SHOT DOWN.

1318 CIVILIANS DIED IN NORTH VIETNAM.

I actually recorded hours and hours of F-4 bombers taking off from Udorn base, a pair every 30 seconds for weeks, with the idea that if I ever got the urge to return to Udorn, I would first listen to that recording to discourage myself.

From the "Legacies of War" website. http://legaciesofwar.org/:

"From 1964 to 1973, the U.S. dropped more than two million tons of ordnance on Laos during 580,000 bombing missions—equal to a planeload of bombs every 8 minutes, 24-hours a day, for 9 years – making Laos the most heavily-bombed country per capita in history. The bombings were part of the U.S. Secret War in Laos to support the Royal Lao Government against the Pathet Lao and to interdict traffic along the Ho Chi Minh Trail. The bombings destroyed many villages and displaced hundreds of thousands of Lao civilians during the nine-year period.

"Up to a third of the bombs dropped did not explode, leaving Laos contaminated with vast quantities of unexploded ordnance (UXO). Over 20,000 people have been killed or injured by UXO in Laos since the bombing ceased. The wounds of war are not only felt in Laos. When the Americans withdrew from Laos in 1973, hundreds of thousands of refugees fled the country, and many of them ultimately resettled in the United States."

Regions in Laos that were bombed are highlighted.

"Here are some other startling facts about the U.S. bombing of Laos and its tragic aftermath:

- Over 270 million cluster bombs were dropped on Laos during the Vietnam War (210 million more bombs than were dropped on Iraq in 1991, 1998 and 2006 combined); up to 80 million did not detonate.
- Nearly 40 years on, less than 1% of these munitions have been destroyed. More than half of all confirmed cluster munitions casualties in the world have occurred in Laos.
- Each year there are now just under 50 new casualties in Laos, down from 310 in 2008. Close to 60% of the accidents result in death, and 40% of the victims are children.
- Between 1993 and 2016, the U.S. contributed on average $4.9M per year for UXO[4] clearance in Laos; the U.S. spent $13.3M per day (in 2013 dollars) for nine years bombing Laos.
- In just ten days of bombing Laos, the U.S. spent $130M (in 2013 dollars), or more than it has spent in clean up over the past 24 years ($118M).

---

4 Unexploded ordnance

69

Departing Thailand

30 December 1972

When it came time for me to depart Udorn my friends Buzz, his wife Phyllis, her son Keith, Gary, Barbara, Howell and his wife Annie, escorted me to the night train to Don Muang International Airport (BKK). We hugged and said our goodbyes. On the steps of the train as it started to move, I turned around, dropped my trousers and mooned – not my friends, they knew it was not meant for them – but Udorn and Thailand and Air America in general. After moving a few feet, the train stopped. Apparently some Thai official had caught my act and stopped the train to find out what *farong* had done such a heinous act. I sat in my seat and watched concerned train conductors search the cars for a naked person. Of course they did not find him, as I had instantly pulled my trousers up, and was now sitting on the evidence. The train soon departed again for my last journey within Thailand.

Buzz shot a photo of my bare bottom on the train steps. He sent the photo off to have a poster made for me, but it got lost in the mail.

Before I departed Udorn, Gary and I made a $50 bet. He bet me that Dennis Kawalek would not marry his Thai girlfriend Dang. A year later he sent me a check. I won again.

31 December 1972
Hong Kong

During the last day of 1972, I caught up with Liz, the exotic Pakistani-Chinese girl from the Up Club toga party. We completed the unfinished personal business between us. Here I re-learned a great truth. Chinese women are a lot like their food: an hour later you want another one.

I brought in the New Year of 1973 at the top of the Apollo Lounge on top of the Hyatt. I watched as the Queen's own Gurkas presented the British colors for the ceremony.

Then I was off to Australia for a month. For tax reasons, I had to kill about four weeks before returning to the U.S. I rented a small holiday apartment on Bondi beach near Sydney and made the most of it. I soon discovered that whenever I went into a bar and ordered a drink, some bloke beside me would ask, "Are you a Yank?" I rarely had to buy a drink. My time in Australia is almost worthy of another book all by itself. I spent a couple of days in Fiji on the way home.

27 JANUARY 1973

THE PARIS PEACE TALKS WERE SIGNED BY BOTH PARTIES.

The war was basically over. All the rest was just winding down the operation and pulling out all the operatives and vital equipment, which took more than a year. Had I hung in there for less than two months more, I would have been able to work without worrying about the Strela missile and other man-made hazards. The Company began laying off pilots and transferring many to Saigon. Gary lived in Saigon for more than a year.

With all the turmoil and change, our pilot's union, Far Eastern Pilot's Association, (FEPA), disbanded. My timing regarding the union was perfect. It had been formed just before I arrived at Udorn and it disbanded right after I left. Despite my many fears of repercussion for my various misdeeds, it never occurred to me to reach out to FEPA for help. The major benefit I received from being a member was that during the negotiations in the formation of the FEPA union, the pilots gained a huge raise in pay.

29 April 1975

That iconic photo of the "Last Helicopter Out of Vietnam" was an Air America Bell. The pilots on that famous Bell were Bob Caron and "Pogo" Hunter. Pogo has since died. I visited with Bob at the Air America Reunion in Tampa, Florida, late May 2016.

If you have enjoyed my book,
I sincerely ask you to go the amazon
web site and write a review. Authors
live and die by their reviews.
- Thank You!

70

More Follow-up Stories

One - My First Air America Reunion

July 1987

I attended my first Air America reunion at Dallas-Fort Worth Airport. I helped move the heavy brass plaque from the truck to the museum at the University of Texas, Dallas in Richardson.

This plaque commemorates the 242 people who died in the line of duty while working for Air America and Civil Air Transport (CAT). These 242 people were not CIA operatives. They did not earn stars on the agency's wall of honor.

Hours later, in the wee hours, several of us were sitting around a booth in the long-closed hotel bar, passing around a bottle of Scotch, swapping war stories. I started a story off stating that I had recently flown a French-built Aerospatiale Allouette III in California on a U.S. Forest Service contract. Before I could continue, one of the other pilots, Willy Peter (a pseudonym to protect the identity of the family and the pilot), sitting across the table interrupted me to blurt out, "You have French time? You want a job?" Curious, I said "Maybe, what you got?" The next words out of his mouth I quote directly:

"How do you feel about killing niggers?"

I was shocked, but he had my attention, so I asked him to tell me details. He was involved in the country that used to be called Surinam and Dutch Guyana before that. It had become the communist People's Democratic Republic of Surinam. He was working for the government in their attempt to persuade the bush Negroes to come in line with the communist government's way of thinking. Their persuasive techniques included bombing, napalming and machine-gunning the villagers – whatever the government thought might be necessary to bring the people around to government control.

The job paid $10,000 a month with alternating months off.

I had visited the bush Negro villages as a tourist in 1968. You couldn't ask for a more peaceful, loving people than those descendants of escaped slaves bound for the southern plantations in the days when cotton was king. NO WAY was I interested in mass murder. Not at any price. I declined the job offer.

Two months later Willy Peter died violently when his helicopter was blown out of the sky by a rocket-propelled grenade.

I had an offer to do some mercenary work during the Iraq-Iran war in the early eighties. A lady friend of a friend was dating a "retired" Army Green Beret colonel. She asked me on his behalf if I would be interested in flying for Iraq in their war against Iran. I was tempted. It was $75,000 a year, "no direct combat, unaccompanied, and good time off." I did not follow up. By then I was married and had two toddlers at home. My priorities had changed.

Two

A Never-met-before Friend from the Far Past

Before I left for U.S. Navy Flight School in 1964, I dated a young lady whom I met on a blind date in San Francisco. We got close and I hated to leave her to go to Pensacola for Navy flight school. We corresponded and I telephoned her every Sunday for six months until I had completed the primary part of flight school. When I came home on leave over Christmas, we spent much time together. Then we stopped communicating and drifted apart.

I remembered that her father worked in Hong Kong. She told me what he did, but it did not make an impression on me at the time. It was many years after I returned to the U.S. that I accidentally met her father. I attended a Marine Corps birthday luncheon in Santa Rosa, California, where I lived at the time. I went expecting to meet up with several Marine friends, but no one I knew showed up. I was a bit lost. A fellow Marine, a WW II veteran of the Battle for Okinawa, invited me to join his table. I sat two seats from the end and met about seven Marines in the group, all WW II vets. The fellow sitting across from me said his name, but it did not at first ring a bell for me as it should have. Later in conversation, he began to ask me lot of pointy question about my Vietnam experience. He seemed so knowledgeable, that I finally asked him, "Were you with the Marine Corps aviation in Vietnam?" He said no, he had been the station manager for Time-Life magazine in Hong Kong during the war. I looked him straight in the eyes and said, "You're Deedee's father!" He said yes, he was. Thirty-four years after I dated his daughter, I met him. She had married someone else and had a wonderful life with him.

My second R&R from Vietnam was to Hong Kong, and I visited several times while flying with Air America. I must have walked past his office near the Hong Kong Island ferry 100 times. I often wondered what might have happened if I had continued to date her or even married his daughter. I took a lot of pictures in Vietnam. Had I shown him the photos upon my return, might he have hired me to be a combat photographer and sent me back to the war?

I like to think that that did not happen because had I returned to the war as a combat photographer, I would have been killed and left his sweet daughter a widow.

Three

1994 – Another Midnight Rambler

In early 1994, I purchased an older automobile, a 1963 American Motors Nash Rambler "American." I thought I might fix it up. I have never been one to get emotionally involved with my cars, but this one was so funky that I began to think about giving it a name. What could I possibly name such an old, plain, funky automobile? I began to think, "What goes with "Rambler?"

Of course! "Midnight." Is it any coincidence that the seats in my new-old car had a lever that allows the front seats to recline all the way down to the level of the back seats, creating a bed that fills the entire passenger compartment of the little buggy?

Author with the Midnight Rambler in 1994
I immediately decided I did not want to fix it up, so I resold it.

Four

A funeral

27 March 2015

I attended the funeral of 93-year-old former Air America pilot, Charlie Stoudt. He died near Spokane, Washington, with 30,000 flight hours in his logbooks. Charlie was truly "an old China hand." He told me that a few years before, he had spent years writing his memoirs, only to lose all his work and all his artifacts in a home fire. He said he did not have the energy to start over. Too sad ... a lot of amazing stories lost.

A few years before Charlie's death, at an Air America reunion in Dayton, Charlie and I had lunch at the hotel bar. The main topics of our conversation were the physical attributes of the beautiful young woman serving us.

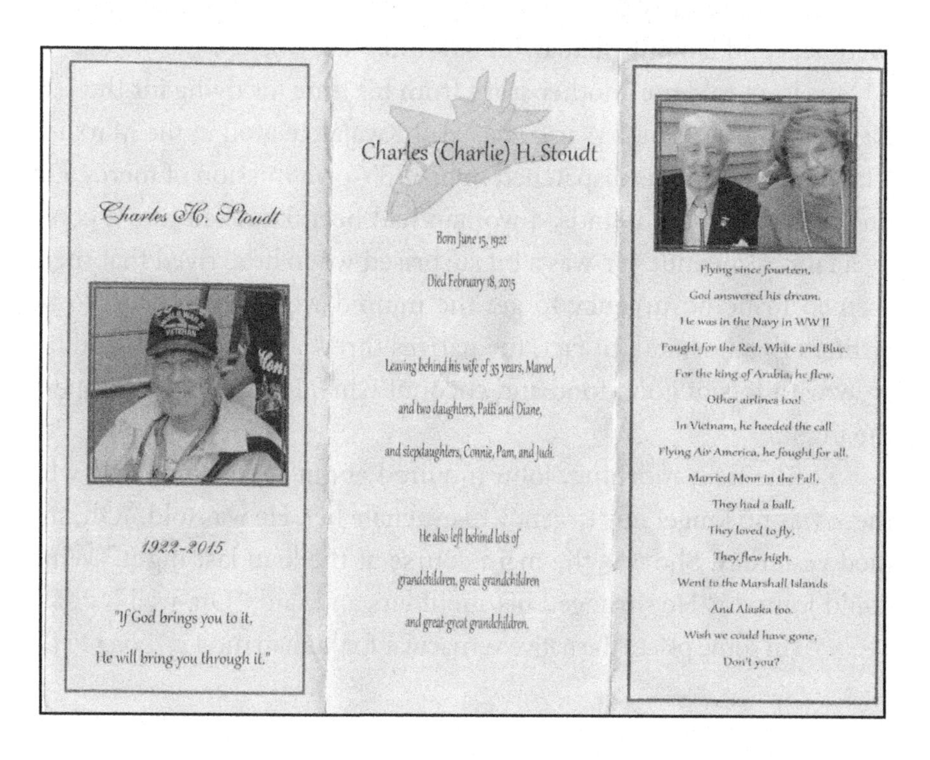

Charles H. Stoudt

1922-2015

"If God brings you to it,
He will bring you through it."

Charles (Charlie) H. Stoudt

Born June 15, 1922

Died February 18, 2015

Leaving behind his wife of 35 years, Marvel,
and two daughters, Patti and Diane,
and stepdaughters, Connie, Pam, and Judi.

He also left behind lots of
grandchildren, great grandchildren
and great-great grandchildren.

Flying since fourteen,
God answered his dream.
He was in the Navy in WW II
Fought for the Red, White and Blue.
For the king of Arabia, he flew,
Other airlines too!
In Vietnam, he heeded the call
Flying Air America, he fought for all.
Married Mom in the Fall,
They had a ball.
They loved to fly,
They flew high.
Went to the Marshall Islands
And Alaska too.
Wish we could have gone,
Don't you?

Former Bell pilot John Greenway was also at Charlie's wake. John was famous for an incident that occurred near Long Tieng in 1972. He landed on a small pad close to a steep slope. Two local radio operators, Hammer and Nail, scrambling to make a hasty exit, did not notice that the steepness of the path that took them close to the rotor blades. They walked right into the blades, turning both their heads into pink mist! Later that day when asked how his day was, John replied, "It was murder out there today."

That is how I heard it in 1972.

At Charlie Stout's funeral, I asked John about that incident. He said, "The two radio operators were clear of the rotor blade tips, when for some unknown reason, my copilot moved the cyclic stick from side to side, lowering the blade tips just enough to catch and obliterate the heads of Hammer and Nail."

I could tell that the event still causes John grief even 45 years later. He did not say the name of the copilot. I am sure that pilot remembers and laments that awful day, too.

John told me another story from his time for flying for the U.S. Navy right after World War II. Based at Kwajalein atoll in the Marshall Island Group, he was dispatched in his PBY on a mission of mercy. On one of the outlying islands, a woman had been knocked into a coma by a falling coconut. He was a bit surprised when he arrived that there seemed to be no urgency to get the injured woman on board to be airlifted to a hospital. In fact, the natives threw a luau for him and his crew with lots of good food and coconut wine. He and his crew spent the night.

The next morning, John inquired about the woman and why there was no longer any urgency to evacuate her. He was told, "Oh, she died yesterday. She was the main course at the luau last night." What could John do? He shrugged his shoulders and said, "Oh, well."

(In some places the native vernacular for human meat is "Long Pig.")

Five

Madras, Oregon

The last weekend of August 2015, I drove over to Madras, Oregon for the "Air Show of the Cascades." Great show – lots of World War II aircraft including a flying B-17 Flying Fortress, a P-38 Lightening and a P-51 Mustang. I rented a booth and sold about 40 of my Vietnam books. There was a group of young ladies dressed as 1940s pin-up girls. I had my picture taken with two of them.

Author with two pin-up girls

Saturday afternoon, when the air show was over, as I carried a box of my display junk out to my car, I found myself walking beside the pin-up girls. I impulsively reached into my shirt pocket and gave one of them a $20 bill as a donation to help their cause. One asked another, "Who's he?" Another said, "Oh, he's the helicopter pilot that was selling his book about helicopter flying." A third girl then said, "My grandfather was a helicopter pilot in Asia. His name was Chauncey...." Before she could finish, I said, "Collard." I flew at Air America with her grandfather, Chauncey Collard.

Chauncey Collard's granddaughter Camille.
She is a beauty and a charmer.

We made a breakfast date for Monday. Camille, her mother Raleigh, her baby girl, and her boyfriend (who was a student at Leading Edge helicopter school in Bend) joined me. We had a nice little Air America reunion right there at Jake's restaurant, downtown Bend.

After arriving back home, I found Chauncey's business card in my card file. In his mid-fifties at the time, he claimed to be the oldest active helicopter pilot in the Vietnam War. I have no doubt he was. No military helicopter pilot would have been flying daily missions at his age.

Air America 1965 - 75
A.C.P., Capt., SIP R/W
Oldest Active Helicopter Pilot
in Vietnam War

I remember Chauncy telling me that he grew up In Sausalito, California, immediately across the Golden Gate from San Francisco. When he was a mere lad, he saw a forest of masts on the bay from the many ships abandoned by their crews rushing to the California gold fields in the 1849 gold rush.

Six

A Final Salute to Good Friends

After Air America, five of us young former Air America helicopter pilots settled in Northern California. We had all known each other from our early days in Navy Flight School through Vietnam and again at Air America.

"Buzz" Baiz returned to his agrarian roots and bought a vineyard near Selma, California. He and Phyllis and Keith grew grapes for the raisin market. He and Koeppe went partners on a second small farm. They did well, mostly on the rapidly increasing values of California farmland. After a couple of years, the partnership was not happy for either of them so they sold the second piece, doubling their money.

I had a job nearby flying on contract for the U.S. Forestry Service in the Sequoia National Forest above Porterville, 7,200 feet above the Central Valley. The five of us would gather at Baiz's farm, drink beer and swap war stories. Our wives would patiently sit and talk among themselves, letting us vent. It was good therapy to share with others who understood what we had been through.

Connolly died 2 July 1975. I cover his death in the dedication.

Howell was learning to sing and play guitar. He wanted to be a rock star. He played at local restaurants and bars in the Porterville, California area. I wrote earlier about the incident when our gunner killed a little girl in her bed. I know that incident, plus whatever else he experienced, caused Steve to suffer from a bad case of PTSD. I know because I discovered later that I also suffer from PTSD. He had all the signs. His sweet wife, Annie, had divorced him.

He was a lost soul. We could tell he was very depressed, unstable, and needed help. But none of us yet knew anything about PTSD, so we just loved him for who he was at the time. I fear he may have gotten involved with drugs and may have been selling them. Then he simply disappeared. No one knows why or where. I called his parents in Oregon about 20 years later, and they had not heard from him for decades. My guess is that he got on the wrong side of some bad people in the area, and he is buried out in the forest somewhere above Porterville. Or he was one of the multitudes of Vietnam vets who have committed suicide.

Koeppe made a small fortune by buying fixer-uppers in La Jolla and reselling the improved properties at a profit. He married Bronwyn, the Australian ex-wife of another young Air America pilot, Steve Bulkley. They had a son, Andrew. About 2004 Koeppe "lost the battle with PTSD" and ended his suffering with a pistol.

Baiz died in January 2014 of lymphoma, yet another victim of Agent Orange.

I am the sole survivor of this group of close friends.

A well-written piece about the journey of life:

"THE TRAIN: At birth we boarded the train and met our parents, and we believe they will always travel by our side. As time goes by, other people will board the train; and they will be significant i.e. our siblings, friends, children, and even the love of your life. However, at some station our parents will step down from the train, leaving us on this journey alone.

"Others will step down over time and leave a permanent vacuum. Some, however, will go so unnoticed that we don't realize they have vacated their seats. This train ride will be full of joy sorrow, fantasy, expectations, hellos, goodbyes, and farewells. Success consists of having a good relationship with all passengers requiring that we give the best of ourselves. The mystery to everyone is: We do not know at which station we ourselves will step down. So, we must live in the best way, love, forgive and offer the best of who we are. It is important to do this because when the time comes for us to step down and leave our seat empty we should leave behind beautiful memories for those who will continue to travel on the train of life.

"I wish you all a joyful journey."

Courtesy of John McDonald, Haverhill, UK,
(Found on facebook 2016)

Seven

A Book that Will Never be Written

During our time at Air America travelling around the world, two stewardesses from a major airline wrote a book, *Coffee, Tea or Me*? about the hilarious and adventuresome times they experienced. They went into detail about the best places to play and meet men, and the best nightclubs to go to in each port. I have a copy of that book in my personal library.

Gary and I came up with the idea of writing the flip side of the story and entitling our book, Coffee, Tea *and* Me. We even made an outline and discussed how we would mirror the girl's experiences from our side. Unfortunately, because of his untimely demise, we never got to write it.

Writing that book with him would have required extensive field research, and it would have been enormously fun.

Eight

About Rescues

Overall, I kept an informal record of rescues. I remember being hauled out of the woods five times by other pilots and that I rescued other pilots seven times. For a while, I was deficient, but then I had a few weeks in Laos where I made several rescues in a short period. My lifetime final tally is roughly seven for five.

Nine

CIA Award

On June 2, 2001, we had a huge Air America reunion in Las Vegas. More than 2,000 association members and guests attended. My daughters both made it. Gary's mother, Ruth and his son, Joey, made it, too.

At our Saturday evening banquet, representative Jim Glerum, (at that time the acting deputy director of the CIA) presented the Civil Air Transport Association and the Air America Association with a Unit Citation Awards recognizing and commemorating our collective service and sacrifice in Asia and Southeast Asia. The CIA presented individual citations recognizing the specific accomplishments and the contributions of Hugh L. Grundy and Robert E. Rousselot.

Each employee of Air America and CAT received this medallion.

The United States of America

CENTRAL INTELLIGENCE AGENCY

UNITED STATES OF AMERICA

Central Intelligence Agency

IN COMMEMORATION

During the hottest days of the Cold War, the aircrews and ground personnel of Civil Air Transport and Air America gave unwavering service to the United States of America in the worldwide battle against communist oppression. Over the course of four decades, the courage, dedication to duty, superior airmanship, and sacrifice of these individuals set standards against which all future covert air operations must be measured. From the mist-shrouded peaks of Tibet, to the black skies of China, to the steaming jungles of Southeast Asia, the legendary men and women of Civil Air Transport and Air America always gave full measure of themselves in the defense of freedom. They did so despite often outdated equipment, hazardous terrain, dangerous weather, enemy fire, and their own government bureaucracy. Their actions speak eloquently of their skill, bravery, loyalty, and faith in themselves, each other, and the United States of America.

George Tenet

Director of Central Intelligence

2 June 2001

The CIA citation. Says nothing. Says everything. Signed by George Tenant. Author's archives.

Appendix A

"CIA Super Pilot Spill the Beans" by Anne Darling
From OUI Magazine, Premier Issue September 1972

I make money off the war.
I'm over here for one thing—the big green.
At least I'm not a hypocrite.

The CIA's Superpilots Spill the Beans

He earns, in a good year, $40,000, tax free. He admits the war in Vietnam made him cynical about the American military, politicians, journalists, Asians and their wars. Noble abstractions don't mean much to him—these days he talks more about "the big green." The American Dream is still in the back of his mind, and his goal is to save enough money to go back home and start a business or buy some income property and live like a normal person. A decent life in the suburbs.

He is a college man who learned to fly a helicopter in the military and survived Vietnam. Now he pilots in Laos, that rugged, damaged country of mountains and rivers and air strikes. His wife and kids live in Udorn, the grotesque base town across the river in Thailand. He is, yes, a professional. A civilian pilot more efficient than the U.S. military pilots in Laos. He flies for Air America. Though it hardly spans the globe, it is one of the largest airlines in the world in number of aircraft it operates.

He is slightly crazy in a way that all Americans who have looked too long at the Southeast Asia horror show are crazy

Illustration by ANDRE VALMONT

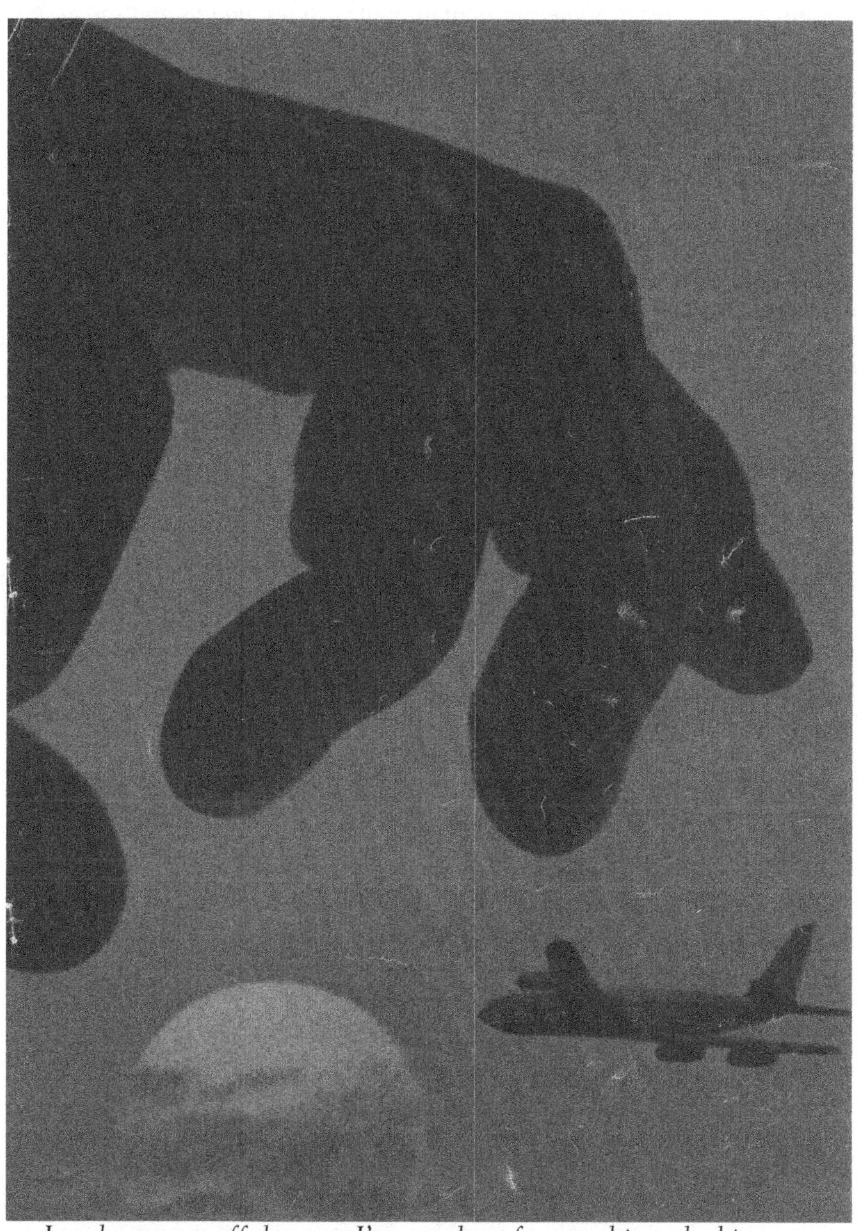

I make money off the war. I'm over here for one thing–the big green. At least I'm not a hypocrite.

Reprinted and retyped with permission.

The CIA's Super Pilots Spill the Beans

They fly the unfriendly skies of Laos, North Vietnam, Thailand and China with cargoes, some say, of opium, prostitutes, bombs and sometimes baby food.

He earns, in a good year, $40,000, tax free. He admits the war in Vietnam made him cynical about the American military, politicians, journalists, Asians and their wars. Noble abstractions don't mean much to him…these days he talks more about "the big green." The American Dream is still in the back of his mind, and his goal is to save enough money to go back home and start a business or buy some income property and live like a normal person. A decent life in the suburbs.

He is a college man who learned to fly a helicopter in the military and survived Vietnam. Now he pilots in Laos, that rugged, damaged country of mountains and rivers and air strikes. His wife and kids live in Udorn, the grotesque base town across the river in Thailand. He is yes a professional. A civilian pilot more efficient than the U.S. military pilots in Laos. He flies for Air America. Though it hardly spans the globe, it is one of the largest airlines in the world in number of aircraft it operates.

He is slightly crazy in a way that Americans who have looked too long at the Southeast Asia horror show are crazy – although not crazy enough to want to stay in Asia forever, as many an old Air America pilot is. The new Air America pilot would have to be the prime selling point for his corporation if his corporation, Air America, Inc.—the ultimate low-profile airline—were the least bit interested in selling what it's got.

Unfortunate, really, that Air America, Inc. does not advertise either its extraordinary customer services or its crackerjack pilots' remarkable record of efficiency—especially when you think of the possibilities for Madison Avenue slogan-makers. "The World's Most Experienced Mercenary Airline." Or—• "Air America: The Wings of the CIA." Or—"Fly the Unfriendly

Skies of Laos —and North Vietnam and China and Thailand and . . ." Then there's the Air America helicopter pilot from Southern California who would be perfect for television spots. A soft-spoken ex-Marine with blond hair, cool green eyes and straight white teeth, he could smile, wink and murmur: "I'm Tommie—Fly Me!"

The first Air America pilot I met was a polite former Marine Corps officer, twice shot down in Vietnam and holding a dozen or so flying medals from the war. It was Sunday in Savannakhet, one of a handful of towns in Laos still called "friendly." On Sundays in Savannakhet, when the war is quiet, a group of American military men and the kind of French expatriates who like to sky-dive from helicopters get together and tote along their parachutes. A Special Forces major, one of those anachronisms in fatigues employed by the United States Embassy in Laos, invited me along for the trip, and the man from Air America flew the chopper.

The day was hot, the sky clear and the season dry. The major, a Bostonian who complained that it was dreary not having smart people to talk to in Savannakhet, recently had progressed from hop 'n' pop (when you hop from a plane, and the static line pops the chute open) to free-fall. Another jumper was a Southern Baptist preacher's son who hoped to make a living on the aerobatics circuit when he got out of the Air Force. The jumps went well, except for that of one Pelletier, a hefty Frenchman who landed very hard on his tail. Afterward, I was treated to a cheeseburger and a piece of apple pie at the Air America hostel, sort of a home away from home for the men when they're flying out of Savannakhet. There the pilot sunned himself, and we made small talk about people who jump out of helicopters when they don't have to —when the choppers aren't burning, that is.

That evening, the pilot and I went to the American Community Association Center, where Americans in

Savannakhet gather to drink, shoot pool, toss darts, see movies and talk to other Americans. We saw a bad film, but not as bad as most of the films that make it to the Americans in Savannakhet. Then we talked about good movies we had missed, the sorry state of journalism in Southeast Asia and Air America in Laos. I wanted to talk about Air America, and the pilot about irresponsible reporters, a gambit that was repeated with nearly every pilot I met over the course of the next two months.

When I ventured to bring up the matter of Air America's connection with the CIA—a most impolitic thing to do—the pilot looked away, as if he hadn't heard me. That, I was to discover, happens often with Air America people. They become hard of hearing, conversations go out of joint, responses are inappropriate; there are changes in tone that say, look you're on the outside, you can't expect me to answer that. But I persisted, rudely, so the helicopter pilot finally answered.

"I signed a contract to fly for Air America, and I don't think of myself as working for the CIA," he said. "The press has accused us of everything from hauling opium to dropping bombs. The other day I read we're flying prostitutes around Laos. It's ridiculous, believe me."

Flying majors from Boston around for Sunday fun was not a typical Air America assignment, the pilot said. A typical assignment was delivering food and mosquito nets to Laotian refugees. "In two years with Air America, I've never been asked to do anything out of the ordinary. I've never even thought of some of these things you're talking about," he said. "The difference between you and me is that you are a curious person, and I'm not a curious person."

Air America is usually described in the American press as "the private airline of the CIA" and by Air America spokesmen as "a private air carrier that serves customers on contract." Of the two descriptions, the latter is probably the more accurate, but both raise more questions than they answer. Indeed, Air America is often referred to as "the largest unknown airline in the world," to which one pilot responded, "and we'd like to stay that way."

Air America has a contract in Laos with the United States Agency for International Development (USAID) that involves mainly the transportation of food and supplies to refugees, who now make up about a fifth of that country's population. Although the company plays up its relief role, several Air America pilots said flying for USAID was the least desirable part of their-work. Pilots are openly critical of USAID personnel and refugee operations; they call the aid workers "naive do-gooders" who squander the American taxpayers' money and line the pockets of local businessmen and corrupt officials. One pilot tells of a project aimed at improving the diet of Meo tribesmen isolated in the mountains of northern Laos:

"For one solid year I flew up there with Gerber's baby food, because some asshole at USAID figured if the Meos would eat a jar of the stuff every day it would be good for them. And for one year I watched those people dump the baby food out on the ground to get the glass jars. There's just no way they're gonna eat strained squash."

Air America also has a contract with the CIA that involves flying secret missions and, in the broadest term, providing combat support for the military effort of the Royal Lao government against the Communist Pathet Lao and their allies, the North Vietnamese and Viet Cong. It is this aspect of their work the pilots, in weekly meetings and memoranda, are constantly warned not to talk about— even to their wives.

One Air America helicopter pilot told me that two years ago most of his flying assignments in Laos were for USAID, but that in the past seven months, as the military situation has deteriorated, he has flown equally as much for "the other customer"—one of the local euphemisms for the CIA. His assignments have included flying cover for Laotian T-28 fighter-bombers, picking up downed Laotian and U. S. military pilots, evacuating wounded soldiers and delivering war materiel. "But it really doesn't make much difference who the customer is anymore," he said, "because the shit is flying everywhere. Laos is getting to be one nice piece of cake."

Air America pilots, however, rarely agree on anything. Some said that the job in Laos was every bit as dangerous as flying with the U. S. military in Vietnam. Others said the opposite. Most pilots said that Air America did not expect them to take the risks the military expected of them in Vietnam; a few complained about being "held back" from missions they wanted to complete. One helicopter pilot said: "I blame a lot of this on the press. Every-time some wild story comes out about Air America, we feel it here.

This escadrille, ranging from old Flying Tigers to recent Marine chopper drivers, calls itself the best goddamned mercenary air force in the world.

We got out of the military because we were fed up with bullshit rules and regulations and reporters riding around with us. We didn't lose that war in Vietnam; the press and the politicians lost it, and now they want to do the same thing in Laos. Air America has held onto this country for three years now. We've gotten our asses shot off doing it. The day I'm told some reporter's gotta sit in that seat next to me, that's the day I'm getting out."

The resentment of Air America pilots toward the press is only slightly more impassioned than the scorn they heap on the Laotians for whom they ostensibly risk their lives. Recently, in Savannakhet, a group of armed Laotian soldiers surrounded the American Community Association Center, where a party had continued past curfew. One pilot said: "Here they are, about to lose their whole damn country, and they come pointing guns at us because we're having a party. We gave 'em 50 bucks to get rid of 'em."

The historical role of Air America is complicated and has only begun to be unscrambled. The company can be traced to its forerunner, Civil Air Transport (CAT), established after the Second World War to operate relief flights and military airlifts in China on behalf of the Nationalists. Several of the original investors in CAT were Chinese, but the founding father was General Claire Chennault, an ardent anti-Communist whose widow, Anna Chennault, has wielded considerable influence in Washington as Taiwan's chief lobbyist. In 1941, before Pearl Harbor, it was Chennault who led the private air force known as the Flying Tigers in support of Chiang Kai-shek against the Japanese. Many of the original CAT pilots were former Flying Tigers. The senior pilot on Air America's payroll was hired in 1946, and five of 23 Air America pilots with 20 years or more seniority are Nationalist Chinese.

Between 1946 and 1954, when the name Air America was first used, the airline performed a variety of intelligence assignments for the U. S. Government in China and Korea. A partial description of those assignments is provided by Peter Dale Scott, a scholar at the University of California, Berkeley, in the book Laos: War and Revolution. Scott stresses the private, quasi-military airline's continuing commitment to "fight Communism" and—summing up Air America's activities—writes: "In calling Air America a paramilitary auxiliary arm, however, it should be stressed that its primary function is logistical; not so much to make war, as to make war possible."

This truculent outfit is incorporated in Delaware as a subsidiary of the Pacific Corporation, whose ownership picture is unclear. Pilots have said the CIA "probably has a piece of the action," along with some of the original Chinese investors and a Hawaiian pineapple millionaire. One man, active in the pilots' union, said: "We'd like to know a little more about the owners ourselves and to take a good look at the company's contracts. We think somebody is raking a lot of money off the top. Maintenance and repairs aren't what they used to be. There's been a lot of talk lately that the company is taking advantage of good pilots."

The air carrier's entry into Laos apparently took place in 1959 on behalf of the CIA. A great deal has yet to be learned about Air America's cooperation with the CIA during the 1960s in Laos, but it is clear that Air America aircraft and pilots were vital in making possible there the CIA's secret military activities. The senior helicopter pilot on Air America's payroll—an ex-Marine hired in the summer of 1960 to fly in Laos, and who is an intense and, shall we say, colorful character who promised to kill me if I used his name in print—called Air America "the best goddamned mercenary air force in the world." Although Air America's operation in Vietnam during the same period involved little more than transporting journalists, visiting dignitaries and entertainers, it's another story in Laos. Air America's short-take-off-and-landing aircraft, helicopters and bulky cargo planes have delivered guns, ammunition, chickens, pigs and even Meo tribesmen. The tribesmen were being trained at Long Cheng, a clandestine CIA installation in northern Laos, to fight Communist soldiers. Long Cheng has been made inaccessible to locally based American journalists because the installation is being used as a staging area for CIA-sponsored spy missions into North Vietnam. Well-trained mountain tribesmen and Thai mercenaries, most of whom speak Vietnamese, are dropped into North Vietnam by unmarked Air America planes piloted by a select few who are paid for the special missions in cash via the "white envelope" system.

Despite all the secrecy, Air America's station manager in the Laotian capital of Vientiane, James Cunningham, Jr., has granted interviews. An affable and reasonably candid former U. S. Air Force pilot, Cunningham spoke at length about Air

America's pilots and how much money they make, its contract with USAID, its aircraft and the hazards of flying in Laos.

Cunningham says a senior helicopter pilot who flies between 800 and 1000 hours a year earns from $40,000 to $50,000, and a fixed-wing pilot earns from $31,000 to $43,000 a year. Salaries are computed by a complicated formula that includes such variables as longevity advances, "hazard-pay" and night differentials. Pilots say they belong to a union—the Far East Pilots' Association. Most conceivable flying situations are covered in the contract's fine print. Pilots are paid for a minimum of 70 air hours a month, so the company sees to it that the men put in at least that much time. The percentage of hazardous flying time varies according to the type of aircraft flown, with the helicopter pilots receiving the greatest allowance for danger. Of the fixed-wing, aircraft, the lower risk is in flying the big C-46, C-123 and C-130 cargo planes; the greater risk is in flying the more vulnerable Porters, the single-engine planes Air America uses on short, dirt airstrips in otherwise inaccessible areas of mountainous northern Laos.

More than 450 men—462 were on the payroll in January 1972, but a few have been killed since then—pilot Air America aircraft. Most of the pilots— about 275—fly in Laos. Fixed-wing aircraft usually are flown by the older pilots with the most seniority, many of whom were in the Air Force or were civilian pilots put to pasture by the commercial airlines and have been flying in Laos for years but with less combat experience than the younger chopper pilots out of Vietnam. Fixed-wing pilots are usually stationed in Vientiane; the helicopter pilots, whose assignments are more dangerous, fly out of the big Thai air base at Udorn. (Thirty pilots also were stationed in Bangkok, Hong Kong, Japan or Taipei, where Air America has its business and maintenance headquarters. Eighty were in Saigon, where the Air America operation is being phased out, and 76 were on leave.)

In recent years, Air America has relied increasingly

upon the more maneuverable choppers because of dangers in flying over Laos—the rugged terrain and seasonal extremes of rain and haze of dust, as well as the possibility of encountering antiaircraft fire. Of those pilots still on the payroll, 100 out of 124 hired since 1968 are flying helicopters. A high percentage of helicopter pilots hired in recent years are former Marine officers because Air America, says station manager Cunningham, appreciates "Marine professionalism." In addition, its helicopter fleet is made up largely of H-34s, the single-engine workhorses Marine pilots continued to fly in Vietnam after they were phased out by other branches of the military in favor of the more powerful helicopters such as the Slicks, Hueys, Cobras, Jolly Greens and Super Jolly Greens—the litany of new whirlybird nicknames.

CIA'S SUPERPILOTS The men who fly for Air America are not modest men; they are often openly critical of regular U. S. Air Force fliers. A A airmen say, for example, that they really earn their pay.

Although there is a good deal of esprit de corps among Air America pilots, who are proud of their status as civilians, there are noticeable differences that are more than generational between two groups—the older fixed-wing pilots who fly out of Vientiane and the helicopter pilots based at Udorn. From my conversation with Cunningham, my impression is that Air America management is partial to the Vietnam veterans.

One chopper pilot said: "The older pilots resent us. We're younger, but we've had more combat experience. We've been to college and they haven't. have in ten. [sic] But I think what really gripes them is that we have good lives a couple of years than the older pilots Some of us have saved more money in to look forward to in the States, and they're stuck with the greatest collection of whores from Seoul to Calcutta for wives. No wonder they can't go home."

Many of the younger pilots see Air America as the best flying job around because it offers them the opportunity to save a lot of money quickly. Several men said they had a goal of $100,000 in the bank or invested, and when they reach that goal, they will quit and go home. Most pilots don't object to being called mercenaries. One man said: "Sure, we're mercenaries. We're civilian mercenaries. We're like a good pro football team. It doesn't need a cheering section to win. And we don't need your American public behind us because we're the first team. We win."

A flight mechanic said he did not mind being called a mercenary but resented Americans back home and journalists who pointed a finger of shame at Air America. "I make money off the war," he said. "But look at your journalists and your television networks and the man who sells the peace buttons and posters. They're all making money off the war. I've got a brother-in-law who criticizes me for 'working for the CIA.' But what does he do? He works with a big defense contract. At least I'm not a hypocrite. I'm over here for one thing—the big green."

The point Air America pilots make about money is that they aren't overpaid. A fixed-wing pilot in Vientiane said: "I've been flying for Air America in Laos for ten years, and in all that time I've never backed up to a paycheck. I've earned every cent of it. We all do." A chopper pilot said: "The Air Force people say we earn too much money, but based on flying time, they get more than we do. They cross that river once or twice a month and pick up combat pay for a couple of hours in Laos. We're out there day after day, 70 hours a month."

Although station manager Cunningham says journalists have exaggerated the rivalry between Air America and the Air Force, his pilots are openly critical of the performance of Air Force pilots in Laos, who supposedly have superior equipment but a much lower rate of assignment completion. A chopper pilot at Udorn told me of a rumor among Air Force people that Air America pilots receive a $1000 bonus for every downed Air Force pilot they pick up. "I see a Jolly Green down," he said, "and I figure he's an American, and I'll stick my neck out any day in the week to pick him up. But I don't get a cent for that, and it burns me up that I can't come back and walk into his Officers' Club to buy a hamburger."

Air America does not seek publicity for its rescues— or for anything else— but its pilots seem to resent the public hoopla when the Air Force accomplishes a rescue mission. A chopper pilot said: "I was a half hour away from some downed airmen one day and couldn't get clearance to go in for them. The Air Force wanted to make its own rescue. Well, it took them 24 hours, and then there was a big story in the papers. They probably put on their monkey suits and threw a party at the Officers Club to celebrate. But if I'd gone in, it would have taken two hours, and nobody would have said a word for Air America."

The pilots who fly for Air America are not modest men. They say they're the best in the world, and it's probably true that they are the best in Laos—for good reason. They have been flying there for years and know every mountain and dirt runway in the country. The Air Force pilots are usually on one-year tours and spend far less time across the River. And the military pilots are allowed mistakes; Air America pilots are not. The company is apparently less concerned with a pilot's mental health than with his ability to perform in the aircraft. One pilot's wife said: "We have a few sickies over here. There's one pilot who likes to beat up on the women he sleeps with. He nearly killed a prostitute last year. A female officer tried to bring charges against him, but it was hushed up. I've talked to my husband about him and he says the company keeps him on for one reason—he's a good pilot." Since 1970, seemingly under pressure, Air America has hired 15 Thai pilots and one Laotian pilot, all regarded by the Americans as inferior in ability. Generally, pilots are scheduled to fly in teams with the senior man in command; the Asian pilots fly either together or with an American who has more seniority.

I recently spent some time in Udorn with an Air America helicopter pilot who had reached the financial goal he had set for himself and was returning to his home in San Diego.

"Air America pilots are no different from any other pilots," he said. "They're all about 60 percent crazy. And then you've got a lot of Marines here, too. A psychiatrist made a study of Marine officers and found that 74 percent of them were mentally unbalanced. Not so that they couldn't function. Just kind of crazy. They need to prove themselves. It had something to do with their mothers."

Another problem, he said, is that I was seeing the pilots at play, not at work. "These guys get their asses shot off six days a week, then come back and get hammered and act like fools.

CIA'S SUPERPILOTS More suspicious of Americans than of foreigners, most of these guys are tough, smart, quick, eager to prove their masculinity, and hot on the stick. They've got to be.

Maybe it's because they are basically egotists. You've got the guy with no education who had some brains, so the military taught him to fly and now he thinks he's pretty smart. Then you've got the guy who's been to college and became a pilot, and he really knows he's smart. So you've got a pretty smart bunch of guys around here."

An Air America pilot's wife with whom I talked on several occasions also discussed the men—but in somewhat more flattering terms. "These men are good at what they do," she said. 'They're professional and in good physical condition. They work hard and they play hard. They're young and strong and their masculinity is important to them. They aren't intellectuals or particularly cultured; but they're smart and quick, and physically, they're an elite group."

Another woman told me the Air America job allowed many pilots to 'prove themselves as men." Her husband, an ex-Marine, had tried once to work for a large corporation where he was told not to wear a yellow shirt or to try any "Marine rough stuff on the employees—they're union." She said: "The jobs in the States, even the flying jobs, are emasculating. Here, I know my husband is taking a lot of risks and I worry about him, but I wouldn't have it any other way. I know how important his manhood is to him."

There have been numerous helicopter crashes, but in the past two years, only some ten Air America pilots have been killed, a remarkable record considering the time they put in and the dangers involved. In nearly every case, the pilots killed were flying single-engine Porters that experienced mechanical failures or crashed into mountains in bad weather. In June, however, Air America lost its only black helicopter pilot. A Vietnam vet from Los Angeles, he was hit in the head by enemy fire while attempting to land on a mountaintop in Laos.

As to frequent reports in the press of clandestine flights into China by Air America pilots for the CIA, sources in Vientiane say those flights were halted before Nixon's visit to China. Other reports have said Air America aircraft were involved in the transportation of opium and heroin out of northern Laos. The implication is that the CIA, using Air America pilots and planes, is —or has been—in the drug business. Air America pilots told me that some of their planes may have been used to transport opium, but that the purpose was not to finance CIA activities. One pilot said: "What people don't understand is that there are thousands of people in Laos who depend on the sale of opium for their living. It's their only cash crop. If the CIA bought opium from them and Air America carried it out, it was to keep it from reaching the world market." In the strange and complicated context of Laos, this explanation seems plausible. Another pilot said: "If you want to know who's transporting opium and heroin out of the country of Laos, I'd suggest you take a look at the Laotian military aircraft, not at Air America's planes."

I read a memorandum sent to helicopter pilots that described a fairly typical Air America assignment for the CIA. It was dated January 19, 1972, and signed by John D. Ford, an Air Ameri https://mail.google.com/mail/u/0/#inbox ca official in Udorn. The memo provided information based on a recent trip by helicopter crews to Changwat Nan—code-named T509—a town in northern Thailand near the Laotian border, where anti-government guerrilla activity has increased substantially in the past year. It gave instructions for future medical evacuation missions and described some problems in radio communications. Security in the area was described as good, "with many soldiers around," but pilots were warned not to leave any valuables in the aircraft, because the Thai soldiers "will nose around at night." The pilots also were apprised of the location of the town's best hotel, a facility without hot water, but for four dollars a night extra, air conditioning was available. The memo alerted pilots that local merchants would not accept U. S. dollars and that Western influence in the area was "negligible." The people, however, were "pleasant' and the food "clean."

Before making my first trip to Udorn, a pilot in Vientiane cautioned me to be careful because "it's another ball game over there, and they don't like people snooping around."

Station manager Cunningham earlier had suggested that I make no attempt to contact his counterpart in Udorn, Clarence J. Abadie, who is "gun shy when it comes to publicity." Abadie, an ex-Marine helicopter pilot, has been with Air America for 12 years but station manager for less than two. He is unpopular with the pilots he supervises. They call him "a very uptight man." Local graffito says "Clarence Abadie reads Ramparts." I attempted to reach Abadie by telephone but couldn't get through to him. Two pilots refused to carry to Abadie's mailbox a letter requesting an interview. One said: "If you know what's good for you, you'll leave that man alone. That place is a hornet's nest."

During nearly a month in Udorn, I visited the air base five times. Thai soldiers guard the entrances and, as a rule, they do not challenge any Westerner attempting to enter. For identification I presented either a U. S. passport or an expired New York State driver's license. A pilot told me that American Express cards also work well. On my first day in the city, I spent two hours talking with pilots in the Air America compound that includes a large hangar, terminal, office buildings, restaurant, bar and swimming pool. A week later, I was asked to leave when I attempted to enter that compound. A pilot, who has since resigned from Air America, was warned about the security implications of assisting me in Udorn. He said: "You've got to understand what security means here. It doesn't have anything to do with the enemy. The North Vietnamese know everything we're doing. They're not the problem. The security Air America is concerned about is being secure from the scrutiny of the American people. –Anne Daring

Appendix B

"The Hill tribes of Laos and Thailand" by Peter Aiken
LOOKEAST **Magazine,** Nov. 1972. p 6-10 with permission from ***LOOKEAST*** Magazine.

Life and Death among the Hill Tribes

by PETER M. AIKEN

Just before sunset I walked to a cleared precipice overlooking the Akha village on the chance that the haze might have broken in spots and I could see the valley below, with the straight black line of road that stretches from Chiangrai to Mae Sae on the Burmese border in northern Thailand. But there was no valley. Only the haze, the red glow of some hillside fires and a lingering smell like that of autumn leaves being burned in the yard of a New England home when it wasn't against the law to do so. In the center of the village on the cracked, sun-baked earth some boys were playing what looked like spinning tops. One boy spun his top in the center of a circle while another with queue bobbing on his shoulders tried to

Akha village

Ma-Yai and son

knock it out with a second top. He missed. Gary, the other foreign guest in the village, after a little coaching soon had one bravely twisting in the center. Then a taller boy, barefoot with ornamented leggings from knee to ankle, made a final wind of the string and took a sidearm throw that sent Gary's top out of sight. The hard, sharp crack of wood on wood and laugh-

6

The chief's casket

ter. The victor was acknowledged and a dog ran out barking at the loser.

It was not hard for me, a foreigner, to live with the hill people. I met tribesmen in the markets around Chiangrai early in the morning. Some spoke Thai. I asked them if I could return to the hills with them. At first they were shy, but when they got to trust me, they were very friendly.

I ate their food and brought some of my own to share: canned goods, sardines, biscuits for the children. I spoke with signs, by smiling, sometimes in Thai. I learned to respect their ability to live where they live with as little as they have, to live in the jungles.

There was more than the usual daily activity today in this village. A chief had died and that night there would be a feast and a wake. The skeletal carcass of a small animal hung from the spirit gate at the entrance to the village. Hopefully the evil spirits would come no further. The Akha are animists; very few have been converted to either Buddhism or Christianity. They believe spirits live in all things, causing and curing sicknesses. Near the spirit gate in this village were three carved figures propped up by stakes. Both male and female, with exagerated sexual organs, they were clearly fertility symbols. They are taboo and may not be touched by outsiders.

An older woman with corncob pipe in mouth and a large sack of straw on her back came into the village leading three brown puppies with wooden harnesses around their necks. The Akha are unique among hill people in that they raise and slaughter dogs for meat. This woman was probably making sure a future dinner did not escape.

We are introduced to Lao-shi, the headman, and he invites us into his house to eat. Tethered outside the door are three bright green parrots that squawk and spread their wings when we approach. Over the door-way are several monkey skulls and we must bend over to avoid hitting them. The walls are thatch, the floor bamboo planks that sag and don't quite meet in places so that if you aren't careful a foot might slip through. Pigs grunt anxiously below. The smell of meat being roasted and giggles from the back room. I step carefully back into the kitchen area. In the fire is a dog, charred and sightless, its tongue swollen and protruding, the body minus its feet. Like roasted marshmallows, the children have the dog's feet stuck on the end of sticks turning them over in the flames. The women are shy and go back to their work.

7

Akha girls

When the food is ready we sit on the floor around a low table. The food is placed in the center and we use chopsticks. The mountain rice is the most delicious variety I have tasted. Chewy like sticky rice and tan-coloured, it is very flavorful. The meat is tough and the sauce extremely hot. When nobody is looking, I drop an unchewable piece of meat through a slot to the pigs below. A short pig fight follows. The vegetable dish looks like grass and leaves mixed up with chilli peppers. It tastes much the same. Bamboo shoots go well with the rice though and we open a can of fish which they put in a soup. Homemade rice wine is poured

8

out in bowls and passed around. The food disappears quickly and our hosts know we are satisfied and grateful.

The coffin in the corner is shaped curiously like a ship about to be launched into the sea. It is carved from a large log and sealed tight. There is a mourning period of around ten days and then the coffin is carried west into the jungle and buried. A smell now of freshly carved wood and resin.

Lao-shi is rolling tobacco into a rough shred of paper and slowly speaking broken Thai.

"He was a good chief but we have many problems lately. We came from Burma twenty-five years ago and have not moved from this hillside. This is not our way. My people have always moved on after a mountain becomes dry."

I thought of the walk up to the village. A crackling sound like an animal stampede in the distance and smoke everywhere. This is the time of year when the hillsides are burnt to prepare for the next year's planting. I wanted to ask the Yao guide to stop because he was leading us to where the flames came near the edge of the path. But he walked on unconcerned and I guessed that the flame would not jump the two foot path. My instincts were telling me to run very quickly in the opposite direction. Without even a backward glance. the guide kept walking on, through and finally above a fire that was clearing the mountain side.

Lao-shi suddenly calls to a child who brings an ember from the fire to light his cigarette and continues.

"We want to move but some people tell us this is not possible. The hills are not empty as they once were. Even the monkeys that we once ate no longer climb in trees here. They tell us to stop growing opium so we tried coffee. We did not make much money from it. Most of our money comes from selling opium. Without it we would have no silver or cloth to wear or guns to hunt with. Our children will all run about naked."

Ma-yai

People are beginning to arrive and activity increases in the back room. Men and boys with their hair in queues, a few with Chinese-looking skullcaps. And the women coming in shyly and quickly going to the back room all wearing tall hats made of split cane, animal fur dyed red and large silver Burmese coins across their foreheads. On the porch a game has begun. Men are tightly packed around what looks like a checker board. They call out numbers and bets and a man sitting on his haunches near the board with a handkerchief wrapped around his head rolls dice across it. The wake has begun and it is a little like the Irish would have it. Inside men are stretched across the floor, some now sleeping and others with their heads propped up on pillows holding pipes over flickering oil lamps. The light dances on the walls and the shadow of the coffin in the corner, shiplike, bobs up and down as if in a choppy sea. And a mild sweet odor drifting up and around; memories of a painless sleep.

In the back room are the women. You can tell the married from the single by the size of their helmets. The single girls wear shorter ones with fewer silver decorations. They may be seen on certain evenings sitting around the courting circle in every village laughing and singing with their suitors. Now they are sitting in a semi-circle facing a long board with holes and designs on it. One girl puts a stick into one of the holes and together they sing a verse that it repeated over and over again as in a chant. Prayers for the dead or a young girl's wishes for the future? I did not ask but watched for a long time. They did not seem to mind.

"They must never take their hats off," Lao-shi told me, "if a man is near. No man may ever look upon his wife without her hat on."

For every hill tribe there are strict rules of dress for the women. Certain articles of clothing must always be worn regardless of heat, rain, cold or wind. The uniform is part of the tribe's custom and the woman's pride. The wealthier she is, the more silver adorns her. Even the very young girls—not old enough for silver—wear hats of colorful flowers while working or playing. Many older women also wear a silver medallion around their necks and some have necklaces.

Ma-yai is sitting near the fire in the back room and speaking Thai with Gary. She is a beautiful woman, married for a year now with one son. She has had some schooling among the Thai people at Mae Chan and knows the ways of both people. She prefers to stay in the mountains.

"It is too hot and dusty in the town." she says. "The dust makes me cough and some people follow and stare at us. Here it is peaceful and I feel happy."

The women here do all the planting and harvesting of rice and other crops. They prepare the food and sew the clothes. The men are hunters occasionally but mainly stay in their houses talking, smoking or sleeping until the hot part of the day is over. The men exist to father children and the women to bear them and feed the village. Ma-yai might have remained in the town but chose not to.

"Once," Ma-yai remembered, "people in this village bowed and paid respect to the full moon. But we no longer do this for we heard men have gone there. And where men go they must eat and leave waste."

So the people here are not totally removed from what is happening in the "civilised world." After learning that the moon had become like an outhouse, I asked Ma-yai about the future.

"Even today they are building a road leading to us. Our children must learn two languages now, the one of their elders and the one of the valley. They must go to school and learn other ways. But they will return to the hills and I hope there will only be peaceful change. Our people do not want to be fenced in but we must learn to use the land better."

The next morning after the wake I woke early and watched the women and children foot-pound the rice. A steady even beat while many of the men still slept. For breakfast we ate more of the mountain rice and some soup. We ate quietly. Soon some young men came in to join the chief and his sons. After the meal they smoked for a while and then rose together. Two lifted the front of the coffin, two the rear, and one on each side. The bamboo floor slats creaked and sagged. It looked light in their hands and they walked down the hillside toward the morning mist still covering the valley. Down into the white-cloud river they disappeared. The tribe had moved from the hills of Burma under that chief, traded around Lashio, and on through the valleys north of Chiengrai burning slopes and hunting wild game. Now here on this hillside they were to stay and live on measured land. Days of nomadic tribal people here and everywhere were quickly coming to an end. The world was becoming too small ●

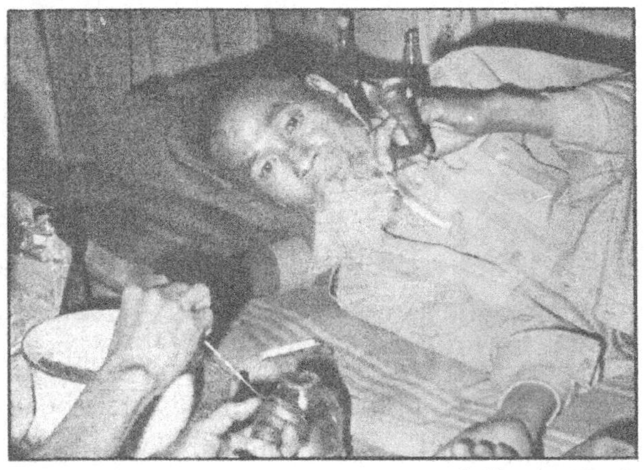

Smoking opium at the wake

10

Article form *Lookeast* Magazine

John Steinbeck wrote this piece to his daughter Alicia in 1967 about Huey pilots, In Vietnam.

"Alicia, I wish I could tell you about these pilots. They make me sick with envy. They ride their vehicles the way a man controls a fine, well-trained quarter horse. They weave along stream beds, rise like swallows to clear trees, they turn and twist and dip like swifts in the evening. I watch their hands and feet on the controls, the delicacy of the coordination reminds me of the sure and seeming slow hands of (Pablo) Casals on the cello. They are truly musician's hands and they play their controls like music and they dance them like ballerinas and they make me jealous because I want so much to do it. Remember your child night dream of perfect flight free and wonderful? It's like that, and sadly I know I never can. My hands are too old and forgetful to take orders from the command center, which speaks of updrafts and side winds, of drift and shift, or ground fire indicated by a tiny puff or flash, or a hit and all these commands must be obeyed by the musicians hands instantly and automatically. I must take my longing out in admiration and the joy of seeing it."

Glossary

AAA	Anti-aircraft artillery
AK-47.	Russian designed combat assault rifle.
Baht	Thai currency, (20 baht to one dollar while I was
there.)	
Baht Bus	Company VW micro bus for transporting crew
members	
BELL	Civilian version of Huey helicopter. 204 and 205
models.	
BIM	H-34 rotor blade pressure indicator.
BKK	Airport designator for Don Muang International Airport, Bangkok.
BX	Base Exchange. U.S. Air Force base general store.
CIA	Central Intelligence Agency
Collective	Control stick in a helicopter that controls blade angle of attack.
Cyclic	Control stick in a helicopter.
ETA	Estimated time of arrival
FAC	Forward Air Controller
FM	Flight Mechanic; many were Filipino. They were
excellent!	
H number	Prefix for H-34 side numbers and radio call signs. Phonetically: Hotel
"hac"	Slang for being in trouble
HAC	Helicopter Aircraft Commander
Helio	STOL aircraft Helio Courier
HKG	Airport designator for Hong Kong International Airport.
ICC	International Control Commission. U N Peace keepers after the war.
ICS	Cockpit intercom
Kip	Unit of Lao currency. (500 kip to one dollar while I was there.)

LPB	Louang Phrabang, Royal Capitol of Laos.
LS-	Lao Site, followed by a number; a large number small airports in Laos.
LS-20A,	aka "The Alternate." Airport code for the secret CIA air base at Long Tieng, Laos,
LZ	Landing Zone
Medevac	Medical Evacuation by helicopter; air ambulance.
NVA	North Vietnamese Army
Oscar Mike	OM, Operations Manager, flight following.
PAA	Pan American World Airlines
PDJ	Plane of Jars-site of large, ancient stone jars of unknown origin.
Porter	Pilatus Porter. Swiss high wing, short takeoff and landing airplane.
POW	Prisoner of war
PTSD	Post-Traumatic Stress Disorder
RPG	Rocket-propelled grenade
RTB	Return to Base
RVN	Republic of Vietnam
SFO	Airport designator for San Francisco International Airport
SKT	Airport designator for Savannakhet, Laos
Strela	Russian hand-held heat-seeking missile.
STO	Scheduled Time Off; Monthly R&R.
STOL	Short Take off and Landing aircraft
TACAN	Navigation radio in aircraft. Tells radial and distance from beacon.
Tango	Tango 08. Royal Thai Air Force Base, Udorn, Thailand.
TIA	Trans International Airlines.
TWA	Trans World Airlines.
TwinPac	H-34 converted to have two turbine engines
Udorn	UTH. Town in North Thailand.
USAID	United States Agency for International Development.

UTH	Airport designator for Udorn, Thailand and Royal Thai Air Force Base
Volpar	Beech 18 converted to turbine engines.
VTE	Airport designator code for Vientiane, the official capitol of Laos.
WW II, 2	World War Two
XW-letters	Registration prefix for Lao registered aircraft.

BIBLIOGRAPHY

1. Aiken Peter, "The Hill tribes of Laos and Thailand" LOOKEAST magazine, November 1972. p 6-10
2. Anderson, Jack Article in San Francisco Examiner, date unknown.
3. Anderson, Jack. 2 ibid
4. Baker, Trudy; Jones, Rachael, Coffee, Tea or Me.
5. Cates, Allen HONOR DENIED
6. Collier, Bill Personal aviator's log books. Several.
7. Collier, Bill Adventures of a Helicopter Pilot, Flying the H-34 in Vietnam for the United States Marine Corps, (available on amazon.com).
8. Connolly, Gary J. Personal Log Books and photo collections.
9. Darling, Anne. OUI Magazine, September, 1972 article: "CIA Super Pilots Spill the Beans"
10. Gonzales, Arturo F. "The Flying Phantoms of Laos" ARGOSY Magazine, February 1963
11. Hamilton-Merritt, Jane TRAGIC MOUNTAINS
12. Karnow, Stanley VIETNAM, "A History, The First Complete Account of..."
13. Lederer, William J. The Ugly American
14. Lee, Patrick Kickers-a novel of the secret war.
15. Leeker, Joe F. PhD. The Aircraft of Air America. http://www.utdallas.edu/library/specialcollections/
16. Morrison, Gail. Hogs Exit
17. Nichols, Stephen L. Air America in Laos, The Flight Mechanics Stories.
18. Robbins, Christopher. Air America.
19. Robbins, Christopher THE RAVENS
20. Shadow government CIA etc.
21. Smith, Felix China Pilot
22. Parker, James E. Jr. Battle for Skyline Ridge
23. Parker, James E. Jr. CODENAME MULE,

24. Parker, James E. Jr. Covert Ops
25. Platt, John Clark, The Laotian Fragments
26. Polifka, Karl L, MEETING STEVE CANYON
27. Readers Digest, ATLAS OF THE WORLD, 1990, for map of Laos p.178.
28. Schanche, Don A. MISTER POP
29. Whittlesley, Peter and Baythong Sinxay, Renaissance of a Lao-Thai Epic Hero.
30. TIME Magazine article, 1972 Sue E.
31. Tobin, James. Ernie Pile's War P152
32. Kotcher, Joann Puffer, VIETNAM Magazine, April, 2002, p50,
33. "Donut Dollies Dodging Bullets."
34. Wolfkill, Grant, REPORTED TO BE ALIVE
35. Yoblanka, Marc Phillip Distant War
36. Yablonka, Marc Tears Across the Mekong

Videos

1. Air America, "Flying Men, Flying Machines" by John Willheim.
2. Air America, "The CIA's Secret Airline."
3. "The Rescue of Raven 1-1," by April Davila. 25-minute video on youtube.
4. "Laos, The Forgotten War" www.youtube.com/watch?v=XB9oXpN2Owg

Web Sites

Air America http://www.air-america.org/index.php/en/

Captain Collier www.captainbillfliesagain.com

H-34 Charlie web site http://dawgdriverforever.blogspot.com/

Helen Murphey's T-28 site http://www.t28trojanfoundation.com/secret-war-in-laos.html

The History Channel, "History Undercover, The CIA Airline"

https://www.youtube.com/watch?v=5qzCJrfTgms

The History Place www.historyplace.com/unitedstates/vietnam

The Legacies of war http://legaciesofwar.org/

The Rescue of Raven 1-1 youtube

The Ravens web site http://www.ravens.org/

USMC Helicopter pilots and aircrew association (popasmoke) site

http://www.popasmoke.com

For a detailed history of Air America, the reader is referred to the books in my bibliography and Air America web site:

http://www.air-america.org/index.php/en/

Acknowledgements.

First and foremost I wholeheartedly thank my wife Carlita for her continued love and support. She was most instrumental in my completing this book. She not only encouraged me to write this book, but without her help to get my life together, I am sure I would be living out in the woods in the Mendocino National Forest in Northern California in a rusty old VW bus sitting on blocks. Her unconditional love and compassion most likely saved my life. She still patiently tolerates an occasional PTSD-fueled outburst.

In addition, I must thank my lovely and charming daughters, April and Summer, who continue to love their old "Pop" even though he was frequently absent from their lives both physically and emotionally. I must also acknowledge their mother, Michele, for her putting up with my PTSD-fired acting out for nearly two decades before finally setting me free. Over the years, she has become a good friend and a wonderful "YaYa" to our four beautiful grandchildren.

I am grateful to my parents, Bill and Emma, and my step-mother, Betty. My brother Cal, who guided me into college-prep classes in high school, provided example and inspiration with his four years in the U.S. Air Force. It was he who inspired me to become an officer. Dr. Albert Heppe taught me well chemistry and physics at Sonoma Valley High School. Sophomore English teacher Beatrice Armstrong awakened my mind to the beauty and structure of the English language.

I am grateful to Captain Thatcher for recruiting me into the MARCAD program, and to my first two flight instructors, First Lieutenant Donny Evans and Captain Barry Schultze, for teaching me primary flying at NAAF Saufley Field near Pensacola, Florida. Without the superb instruction provided by these men, I would have never survived Vietnam and proceeded on to Laos.

I am especially grateful to the crew at Keokee Publishing in Sandpoint, Idaho for helping me through the details of layout, formatting, converting and otherwise noodling this book to bring it to print. Thank you Chris Bessler, Beth Hawkins and Laura Wahl for your wonderful graphics, and all the others at the shop for their parts in bringing this work to completion.

I am enormously grateful to the owner of **oui** Magazine, Ms. Susan Traub of New York City, New York. She most generously gave me permission to use the "Super Pilots" article and the cover of the October, 1972 issue of OUI for this book. The article not only added great flavor to this tale, it inspired the title for the book. Thank you Susan!

A great big thank you to my beta readers, Professor Sylvia Davis, Captain Marion Sturkey, Captain Mark Garrison, April Davila, Customer James "Mule" Parker, Lt. Col. Rich Faletto, Lt. Col. Henry Zeybel, U.S. Coast Guard Captain Tom Beard, all who gave me wonderful feed-back and constructive criticism for this work

A great big thank you to retired Santa Rosa, California fire captain, Ted Boothroyd for his eagle-eye on the last few go-rounds of editing. Without Ted, I would still be straining my eyes looking for more mistakes to correct.

Last, but far from least, a giant burst of gratitude to fellow writer and editor Bonnie McDade who did the final edit and found mak mak more items that needed attention.

And to Ann, Annie, Barb, Barbara, Christine, Crystal, Cynthia, Daryll, Danielle, Diana, Diane 1, Diane 2, Dory, Emily, Fran, Ginger, Ginnie, Glynnis, Grace, Heidi, Helen, Jane, Janet, Judy, Kathi, Kathy1, Kathy2, Kathy3, Kitty, Linda1, Linda2, Linda3, Lyn, Lynn, Liz, Mandy, Marianne, Marie, Marilyn, Margaret, Marsha, Marty, Mary, Molly, Mona, Nancy, Pam, Patti, Petra, Phyllis, Rita, Shelli, Sue, Susan, Syn, Tana, Tani, Teresa, Tish, Toni, Ursula and Vikki … thanks for the memories!

H-34 "Charlie"

Sandpoint Chapter 890 of the Vietnam Veterans of America acquired an old derelict H-34. Captain Bill and his veteran friends have refurbished it enough to make it towable for Fourth of July parades and other veterans' activities. To learn more, check out the H-34 Charlie blog at: http://dawgdriverforever.blogspot.com/.

H-34 Charlie when we first encountered her in Everett, WA. February 2011

H-34 Charlie in May, 2015
She is well beyond ever flying again.

Captain Bill manages this V V A project.

About the Author

After surviving the Vietnam War and Air America, Captain Bill Collier continued to have many adventures flying helicopters commercially all around the world for another 25 years. He flew almost every size and shape helicopter produced in the U.S. from the smallest "whirlybird" to the giant CH-53 "Jolly Green." His last flying position was as senior pilot for the Orange County Fire Department in Southern California, flying the UH-1 "Super Huey." He also flew the French-built Alouette III and the Kaman H-43 "Husky" Syncropter. He likes to report that in all his flying, he has logged an equal number of landings and takeoffs, and never left an aircraft in the sky.

MUG SHOT BY PAUL SATREN, COEUR D'ALENE, IDAHO

In 2008, Captain Bill retired to the small, quiet, peaceful town of Sandpoint in the Idaho panhandle. He keeps himself busy writing about his 32 years of helicopter flying which took him to the far reaches of Alaska, including the farthest out Aleutian Islands, the South Pacific to Kwajalein Atoll, Saudi Arabia, all over the Western U.S., and Tennessee.

He is a life member of many veteran's organizations.

Watch for more "Adventures of a Helicopter Pilot" books, *Alaska,* and the collection of world-roving civilian flying stories … and perhaps a novel.

Be sure to read Captain Collier's first book about flying for the USMC in Vietnam.

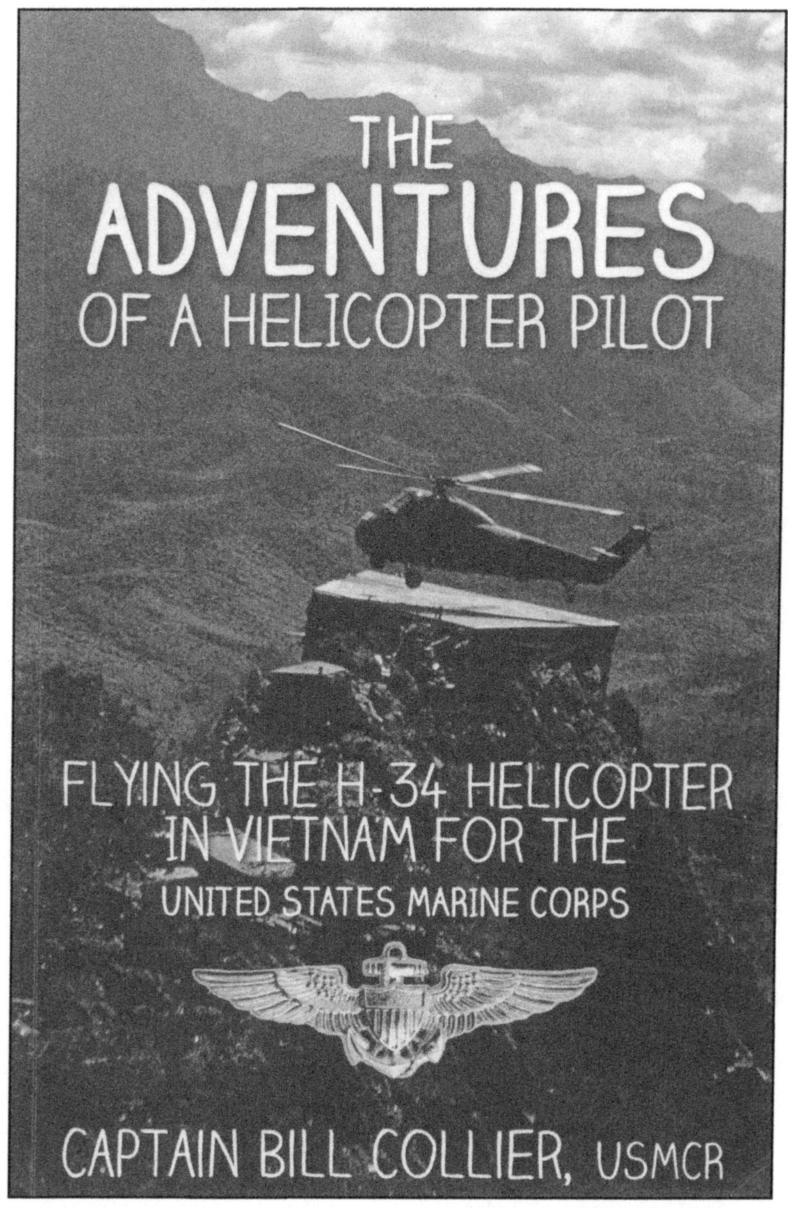

Available at amazon.com as a paperback or an e-book.

Made in United States
Orlando, FL
23 December 2023

41611078R00183